GW01607340

JAGUAR

Philip Porter

JAGUAR

The Complete Illustrated History

FREDERICK WARNE

CONTENTS

Acknowledgements	**4**
Foreword by Sir William Lyons	**5**
Introduction	**5**
1 The Swallow	**6**
2 The SS Marque	**18**
3 The SS Jaguar	**28**
4 The War and After	**38**
5 The XK Engine	**46**
6 The XK Evolves	**66**
Jaguar in colour	*Between pages 80–1*
7 The E Arrives	**82**
8 The XJ Range	**98**
9 Jaguar Competes	**110**
10 Styling Themes	**154**
Afterword by John Egan	**159**
Index	**159**

ACKNOWLEDGEMENTS

As this is my first book I have particularly appreciated the kind and generous help given to me by a number of people.

Firstly I should like to record my gratitude to my old friend Nick Baldwin whose idea it originally was and who helped throughout with advice, photographs and moral support. Jeff Groman of Warnes has shown much patience and assisted greatly.

Regarding photographs Bob Taylor, Bertie Bradnack, David Grounds, Jack Fairman, Mrs. J. B. Parker, Mrs. Bayne, Mrs. E. Nicholson, Mrs. E. Simms, Mr. R. Gore, Don Smith and Mrs. M. W. F. Longlands have all very kindly lent me examples from their private collections. John Owen, Walter Gibbs in his role as Midland Automobile Club Archivist, and Mrs. Gill Hughes latterly of P. J. Evans have all given valued assistance; and Hugh Conway, Gil Mond and Geoff Walker have all helped with photographs and information. With regard to copying of ancient prints, et cetera, Ian Kerr and Graham Willey have been most helpful. Terry Moore of Phoenix Engineering has been a mine of information, and Jeremy Wade and Keith Stewart have both given me the benefit of their advice.

Above all I sincerely thank Alan Hodge, Steve Gilhooly, Norman Dewis and Roger Clinkscales for the tremendous part that they have played in the production of this volume. Alan has as ever been his helpful self, with advice, photos, information and introductions. My old chum Gil has helped enormously checking the text for blunders and bringing his encyclopaedic Jaguar knowledge to my advantage. Norman spent hours talking to me, indeed fascinating me with recollections from the fifties and sixties and lending me a large part of his collection of photographs. Without Roger this book could not have been produced. He has responsibility for Jaguar's own collection of photographs, apart from being kept very busy as head of the Photographic Department. In spite of his ever heavy workload he has helped me enormously, dealing with my many, and I am sure, tedious requests with patience and alacrity. Above all I have greatly appreciated the trust he has shown in lending me many irreplaceable glass negatives.

Finally, I am mindful of the honour accorded me by both Sir William Lyons and Mr. John Egan penning a piece for my modest offering. To them and their many other colleagues at Jaguar, my sincere thanks.

Philip Porter

FOREWORD

by
Sir William Lyons, RDI
President of Jaguar Cars Ltd

I was very pleased to be asked by Mr. Porter to write this Foreword for this new book. It has given me great pleasure to see the enthusiasm of the Jaguar Drivers' Club grow, and my colleagues at Jaguar Cars Limited today will always be as pleased as I am to know that our efforts have resulted in such a strong following for the product, both new and old.

I am sure all readers of this book will appreciate their cars all the more for this insight into Jaguars now no longer produced, but current models will now perhaps be even more enjoyed as the Jaguar pedigree progresses.

INTRODUCTION

This book is not intended to be the definitive volume on the subject of Jaguar, nor is it intended to be a mere coffee table edition. Rather it is meant to be a reasonably comprehensive history of the famous company that began as Swallow, became SS, and then achieved worldwide success, respect and fame as Jaguar.

The dramatic development of the famous Jaguar company from its humble origins as a manufacturer of sidecars through to the present position of automotive eminence worldwide is told chronologically by model. Personal recollections of Jaguar personalities are quoted, adding a human – and sometimes humorous – slant to the history.

I have treated the competition story separately from the history of the production models, believing that to combine both would be to cloud each with too much detail. Treated separately, one can clearly see the development of each, but one should always remember the beneficial effects competitive participation undoubtedly had upon the development of the road cars. Indeed, every Jaguar has had sporting pretensions, some more pronounced than others, and it was (and still is) essential that they took on the opposition in the sporting arenas of the world. That they did so, and with such success in the 1950s, created a foundation on which they have built to this day.

At the conclusion of the book there is a brief chapter on styling influences. Styling is such an emotive and subjective issue that everyone will have their own views and my ideas may be seen by some to be controversial. However, their purpose is simply to initiate discussion.

A very special feature of the book is the over 300 illustrations, almost all period shots that have not been seen before. It is becoming extremely difficult to find Swallow, SS and Jaguar photographs previously unseen and in this area I have been very fortunate to have been assisted greatly by the Jaguar company and several private individuals.

When I was first asked to write this book, the prospects for Jaguar looked dismal. There was a succession of fuel crises, poor labour relations, slipping standards and seemingly little direction, all of them threatening the extinction of a company living off its past. Now, happily, things look a lot brighter, and Jaguar have regained their self-respect and restored their name to worldwide prominence.

As the advertising slogan has it: 'the legend grows'. And this book charts the background and foundation of that legend, a legend created by good design, aggressive business leadership, styling flair, value for money and competitive success – all creating the evocative name of 'Jaguar'.

1 THE SWALLOW

From humble beginnings in 1922 as manufacturers of sidecars Lyons and his partner's newly formed business progressed to rebodying motor cars. Commencing with the ubiquitous Austin 7 the treatment was meted out to a series of cars, the most significant of which was the Standard. Meanwhile, sidecar production continued apace, being a useful bread-and-butter line.

1

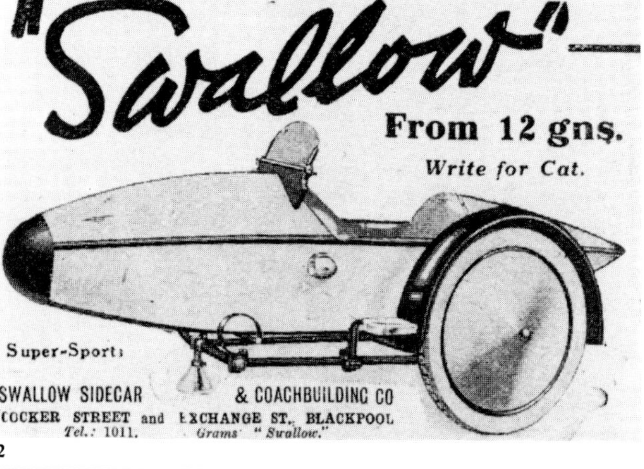

2

3

The great story of Jaguar, and Sir William Lyons, begins with the name Swallow. For in 1922, William Lyons (as he then was) and William Walmsley formed the Swallow Sidecar Company, with premises in Bloomfield Street, Blackpool. Young Lyons had been a motorcycle enthusiast for a while and he added to this a certain eye for style and a head for business, both of which can be detected all the way through this story. It is likely that if this were a novel, it would be dismissed as a fairy tale, because from these humble beginnings Lyons built one of the greatest names the world of motoring has yet known, and is ever likely to know. In the field of growth, excellence, exports and international competitive events, Lyons's company was to achieve prominent success.

For the first 50 or so years this is the story of one man, a decisive, autocratic leader who by a combination of flair, acumen and downright hard work achieved so much. He was Sir William Lyons, who might be described as 'the man who has stolen more Motor Shows than any other'.

He started with just £1,000 and a few men and lads, and when he retired he left behind an empire, including such dis-

1 *A Swallow Model 4 Super Sports sidecar, reminiscent of a Zeppelin (J C)* **2** *A 1928 Swallow sidecar advertisement (J C)* **3** *A racing 'chair' but still with Swallow's usual attention to style and finish (J C)* **4** *The Light Weight de Luxe, described by Swallow as the 'Sidecar for the Connoisseur' (J C)* **5** *This advertisement marked the beginning of the long relationship with Henlys, although their artist appears to have adopted considerable licence in the top illustration (J C)*

The SWALLOW HAS ARRIVED!

SPORTS BODIES for the Connoisseur

The perfection of body-building art developed upon new and better lines. Grace, exquisite finish, greater comfort and better protection, feature in all Swallow products.

HENLYS have been appointed Sole Distributors for the Southern Counties (excepting Kent), and are offering these striking bodies on specially tuned Austin 7 and Morris Chassis at the following attractive prices.

A fine selection of beautiful colour schemes is available. The features include cowl over radiator, Vee screen, draught-proof side curtains or coupe head, etc. For early delivery orders should be placed now.

These bodies transform the Austin into a real car—giving big car comfort and lines. The bodies on the Cowley Chassis introduce an entirely new Morris model.

The MORRIS Cowley Model

The AUSTIN 7 Swallow

PRICES:

	Swallow Austin	Swallow Morris
With cape hood only	£175	£220
With coupe saloon head only ...	£185	£230
With cape hood and interchangeable coupe saloon head	£190	£235

Send at once for Booklet giving full specifications or call and see these marvellous bodies at

HENLYS

SOLE SWALLOW DISTRIBUTORS FOR SOUTHERN ENGLAND (EXCEPT KENT)

1, 3 & 5, Peter St., Manchester Telephone: Central 1780

91, 155-157, GT. PORTLAND ST., W.1. Phone: Langham 3341 (10 lines).

DEVONSHIRE HOUSE, PICCADILLY. Phone: Grosvenor 2271

Service Station: Hawley Crescent, Camden Town, N.W 1 Hampstead 5177.

5

tinguished names as Daimler, Coventry Climax, Guy and Meadows, but above all Jaguar, which became a household name. It is a name synonymous with speed, style and quality.

A man of whom the motoring press and his competitors continually asked, at the announcement of each new model, 'How does he do it at the price?', never really had to sell a car – his order books were always full.

To his credit, he chose and employed some brilliant men whom he moulded into a great team. He understood the significance of competition success and realized the potential sales market across the Atlantic and proceeded to exploit it. Indeed, *The Motor* was moved to comment in its editorial leader of 16th November 1949, following the announcement of the XK120, its achievement of 132 mph at Jabbeke and its victory at Silverstone, that: 'There is, surely, a lesson which transcends this immediate case of a high performance two-seater. It is that the export market will respond to a car having outstanding quality in its class and sold at a reasonable price. The average car is palpably more difficult to sell and it is evident that British designers must be guided by the motto "Aim Higher", and manufacturers must remember when competing with foreign makes the advice given by an admiral to a junior officer "My boy, if argument becomes heated about politics after dinner, and someone throws a glass of wine in your face, don't throw your wine in his: throw the decanter stopper." In the realm of highspeed motoring it is evident that Mr Lyons of Jaguar Cars has thrown a very heavy stopper very hard indeed.'

William Lyons was to throw many heavy stoppers in his long career. Just before his 21st birthday, William Lyons met William Walmsley, a man some nine years his senior. Walmsley was producing, at the rate of one per week from a brick outhouse behind his parents' home, a rather striking sidecar which he was fitting to war-surplus Triumph motorcycles he had reconditioned. Young Lyons detected a commercial possibility with these sidecars which brought style to what had previously been ugly attachments, if production could be raised to a more businesslike output.

Although they had to wait several weeks for Lyons to come of age, Walmsley and Lyons formed the Swallow Sidecar Co on 4th September 1922, with a bank overdraft of £1,000 guaranteed by their respective fathers. Initially they operated from the first and second floors of an old building in Bloomfield

1

Road, Blackpool with several men and a boy, although their workforce quickly expanded and further premises were acquired.

The new company pioneered the use of aluminium panelling and had their own stand at the 1923 Motor Cycle Show. There they achieved the distinction of having their 'chair' displayed on the stands of such august names as Brough Superior, Dot and Coventry Eagle.

A range of different models was offered, with a touring model costing £22 10s, and the most prestigious model listed, the Coupé de Luxe, at an extravagant £30. The orders flowed in, production was raised, the workforce was increased to 30 and another move became necessary, this time to Cocker Street in 1926. These premises were purchased by Walmsley Snr and leased to the company.

Demand and therefore production continued to grow, and about this time Lyons started to export – a field that was to assume greater and greater importance as the years rolled by. Meanwhile, in 1927, another exciting development attracted everyone's attention.

Herbert Austin, fighting the Depression of the 1920s, had designed and introduced the legendary Austin Seven. As is well known, the car was intended to be cheap, economical and reliable to run, and easy to drive, thus bringing motoring to the masses. Lyons and Walmsley may have feared that this new concept might compete with their staple product, or maybe they just saw an opportunity for expansion. Whichever way it was, they grasped the initiative and launched a re-bodied version of the urbane Austin Seven.

There is a definite parallel between the motorcars of the 1920s and those of recent years. As in the 1980s when mass-producers compete to make visually duller and duller boxlike cars, so in the 1920s the majority of saloons were bodily austere and unadventurous. As in recent years specialist manufacturers who innovate and dare to be a little exciting achieve a vogue, big companies

1 *This 1929 model Austin Swallow Saloon shows the standard two-tone design, and it's easy to see why the bonnet colour scheme was referred to as the pen-nib.* **2** *A 1929 model of an Austin Swallow Two-Seater with the cape hood (*J C*)* **3** *A rear view shows the neatly rounded tail so characteristic of these models – note the open vent (*J C*)* **4** *Apart from the Austin Two-Seater, this factory advertisement illustrates the shortlived Morris Swallow Sports (*J C*)*

4

desperately market multitudinous options to attempt to disguise the uniformity of their products, and owners cherish quite bad cars from the past merely because they are different, so in the twenties there was a demand from people who wished to be a little individual; from those who wanted a car with some visual personality and modest elegance.

The Swallow range of re-bodied cars was created to satisfy this demand. Lyons had again tried to bring style to the commonplace. The car had a delightful body constructed of heavy gauge aluminium over an ash frame and a wooden floor. A few early models had bicycle-type front wings that turned with the steering, and no running boards. But fuller wings and running boards were soon added, as you can see from the illustrations. The vulnerable wasp-shaped or beetle-shaped tail was aluminium but with no internal framework. The rear number plate was carried on brackets below, to which on the nearside, the two-inch copper-extended exhaust pipe was affixed. Some tails had professionally built hatches or lids for easy access, but this was not standard. A bullnose radiator graced the front. This item was originally aluminium, then nickel-plated steel and was finally chrome plated. The radiator shell carried the usual wings motif and a honeycomb grille on which was mounted centrally 'Austin Swallow' in script. An aluminium four-piece bonnet, centrally hinged, had 14 fluted vertical louvres on each side for engine ventilation.

The Austin Seven chassis was modified only by the addition on both sides of an angle iron, approximately the length of the running boards, upon which the main body timbers were bolted. A contemporary magazine noted that 'the brakes were as good as might be expected on so small a car, and when used in conjunction (hand and foot) could be termed efficient'!

A neat detachable hard-top (surely one of the earliest examples of this item, if not the earliest) known as the Coupé Saloon Head, could be hinged for easier access – a boon, surely, because the inside width of the body was a mere 38 inches. But after a while this was found to be impractical and the hard-top was merely bolted on.

In spite of these apparently constricting dimensions, Swallow felt able to state in one of their advertisements, 'It combines in ideal form the sweeping lines and performance of the sports car with the luxurious, roomy comfort of the big car – no wonder it is such a popular success'. This was a good example of the company's advertising prose, showing absolutely no regard for modesty, and which was to be a continuing and fascinating theme in all factory brochures and advertisements. While the prewar period was renowned for its flowery promotional prose it seems that the company was pre-eminent even in this sphere.

Announced in 1927 some three months after the Austin, and produced into 1928, was the Swallow-bodied version of the Morris Cowley chassis with 1550 cc engine, known as the Morris Swallow Two-seater. Comparatively little is known about this particular model and it is thought that only a few were made. One theory to account for this was its resemblance to the current MG, which it bettered in price but could not compare with in performance. With Ace discs disguising the artillery wheels, the price was £210, or £220 with wire wheels.

Externally, apart from being larger than the Austin, the body was less curvaceous, with a high bonnet line, V-shaped windscreen, side-mounted spare wheel, and sloping tail incorporating a dickey seat.

To meet its changing role, the com-

1

2

pany title was expanded in 1927 to read 'The Swallow Sidecar and Coachbuilding Co.' About that time the first overseas agent, one Emil Frey of Zurich, was appointed, initially for the sale of sidecars, the production of which continued to expand. In 1982 Emil Frey, at the age of 83, had bestowed on him an honorary CBE for his services to British exports.

In November 1928 the open two-seater version of the trusty Seven was supplemented by the addition of a saloon based on the same chassis.

The company's catalogue stated that, 'The Sports Saloon, embodying the refinement and exclusive distinction characteristic to the luxury car, makes a strong appeal to those of discriminating taste to whom such qualities are essential. Grace of outline, perfect proportions, and exquisite colours, enhanced by seating accommodation for two adults and two or three juvenile passengers'.

The body, again aluminium over ash, had two doors with vertically sliding windows balanced by an internal spring-loaded roller. The window could be held in any position by a central locking lever. The roof was of wooden framework constructed so that in later models a sunshine roof could be fitted to order. Otherwise the top was covered with a black leatherette material. In the roof behind the rear passenger was a square four-inch air vent. The earlier saloons had solid wooden rear floorboards that produced in the passengers a 'knees

Foleshill in 1929, showing the creation of the Austin Swallows: **1** *Austin Saloon beginning to take shape with the construction of its wooden frame.* **2** *Part of the body shop.* **3** *Body frames, now panelled, ready for the paint-shop.* **4** *In the trim shop.* **5** *Now painted, exterior fittings begin to be added.* **6** *Nearing completion, to the left is the distinctive wasp-shaped tail of the Two-seater.* **7** *Completed, the little Swallows await delivery (*J C*)*

6

7

under their chins' attitude. Later, a metal well was fitted with a central hump to house the propellor shaft. The roof forepeak was curved forward, projecting by four inches in front of the windscreen centre V point and $1\frac{1}{2}$ inches at the side posts.

The 'exquisite' colour schemes were, to say the least, bold for the depressed 1920s. The catalogue listed the following 'exceptionally highly polished: Cream and Crimson; Grey and Green; Light Mole Brown and Deep Suede Brown; Ivory and Black; Cherry Red and Maroon; Sky Blue and Danish Blue; Cream and Violet; Birch Grey and Battleship Grey; Ivory and Dark Blue. The first mentioned colour in each case alludes to the body below the waistline; the latter to the wings, wheels, chassis and head above the waistline'.

And if that did not convince the ladies who were attracted to the Swallow Seven, then this further extract from the sales catalogue probably did: 'THE INTERIOR FURNISHINGS are artistically finished in polished mahogany or walnut, the instrument fascia board being neatly mounted with switchbox, ammeter, speedometer, and oil pressure gauge. Incorporated in the design is an artistically designed locker, providing Ladies Companion Set. The floor coverings are of pile carpet, in colours to match the upholstery. An interior electric light with tumbler switch is fitted into the roof, and an adjustable mirror provides clear vision through the rear window light.'

As to sales, Parkers had been appointed distributors for Manchester, and Brown and Mallalieu for Blackpool, when Lyons decided to take one of his new Austin Swallows down to visit the comparatively new firm of Henlys in London. To his surprise and delight Henlys said that they would take 500 provided they could be the sole dis-

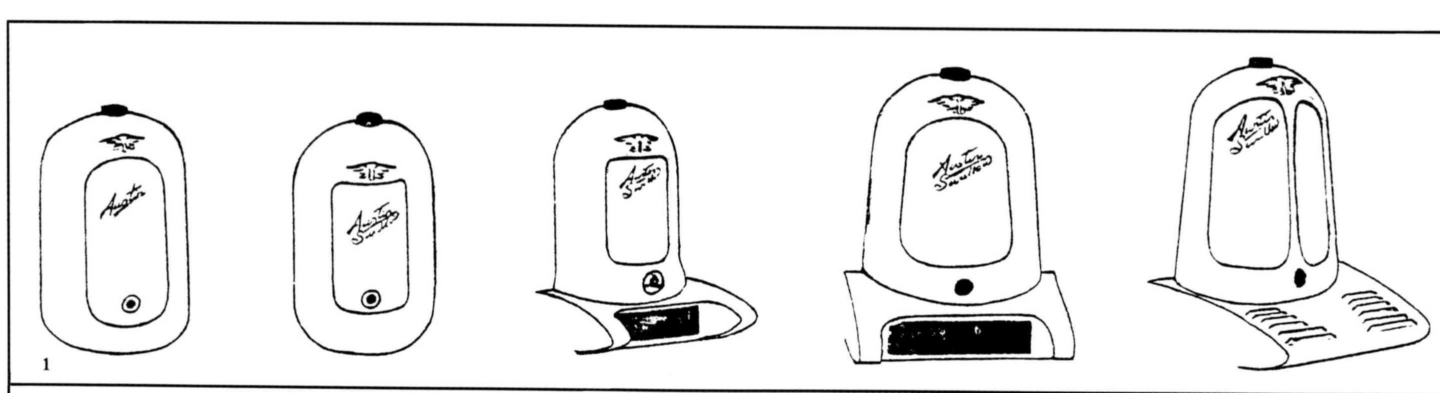

1

2

tributors for the south of England. To this Lyons agreed and set off back wondering how on earth he was going to build the high number of 500 cars.

For this reason, and because the sidecar production was still expanding, another move became necessary. A supply of suitably skilled workers was the overriding factor, so Coventry, being the heart of the motor industry, was chosen, and there they would have the added advantage of being nearer to their suppliers. An old munitions factory at Holbrook Lane, Foleshill, was found to be suitable and the move was made. It was a brave move in 1928, when many companies were going to the wall. Ironically, the Depression probably helped Swallow rather than hindered it, for those who had been used to expensive individually coachbuilt cars found the Swallow within even their impoverished means and, more to the point, it seemed less of a come-down.

In 1930 the Mark II version of the Swallow Seven appeared at the Motor Show with minor revisions. The radiator shell was less bull-nosed and more V-shaped, and had a vertical chrome strip down the centre. The bonnet, an inch-and-a-half longer, now boasted 30 louvres on each side, and most Mark II models had front and rear bumpers. Upholstery was of real leather, interior trimmings were plated throughout and the furnishings were of polished mahogany. Further additions included adjust-

1 *Evolution of the Austin Swallow radiator, 1927–1932 (*Swallow Register*)* 2 *The Swallow treatment worked equally happily on the larger Fiat chassis (*J C*)* 3 *A fine example of Henlys' audacious advertising style, featuring a Standard 9 after Swallow treatment.* 4 *A badge evolved for the Standard, with perhaps a hint of what was to come.* 5 *Another illustration of Henlys' 'modesty' when applied to other Swallow products (*J C*)*

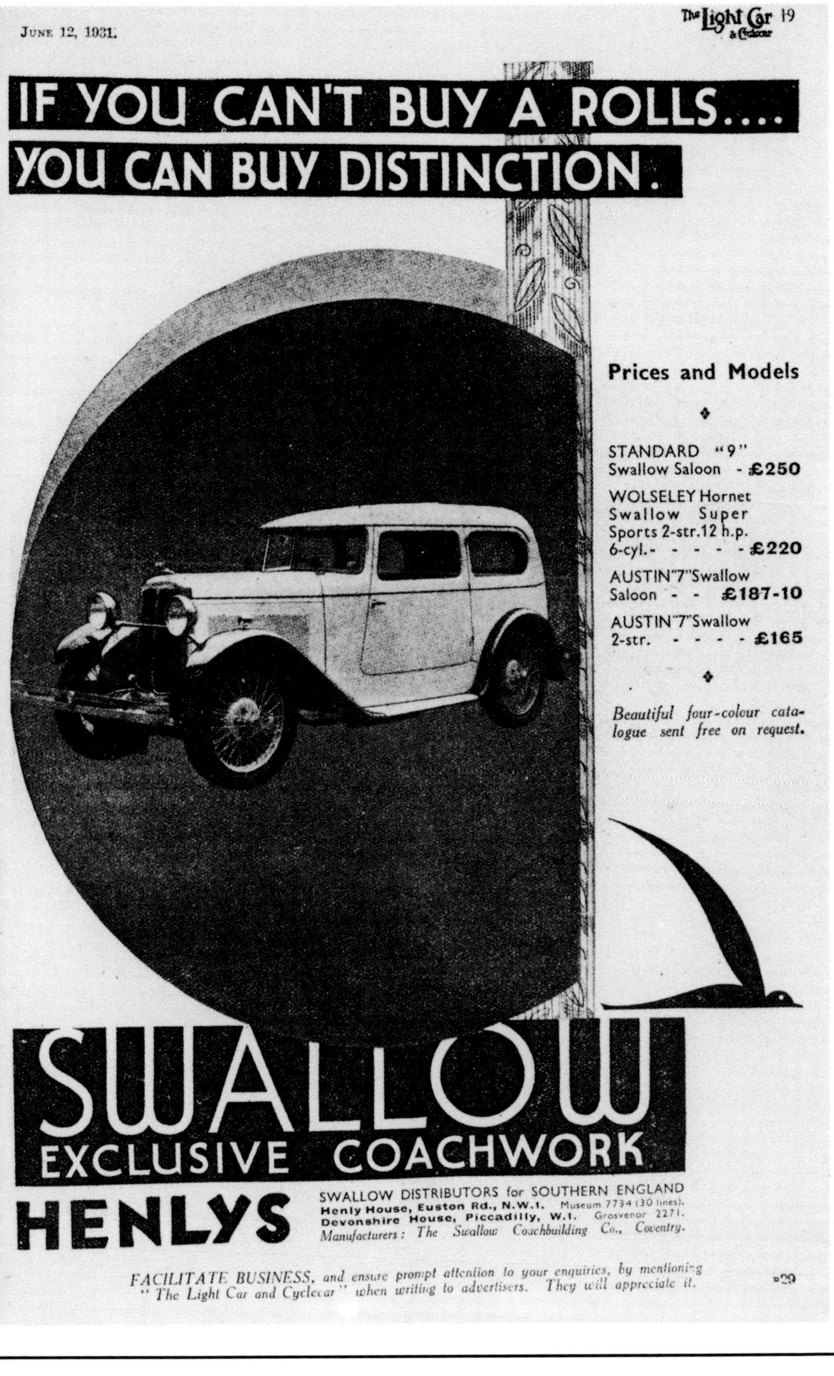
June 12, 1931.
The Light Car & Cyclecar 19
IF YOU CAN'T BUY A ROLLS....
YOU CAN BUY DISTINCTION.
Prices and Models
STANDARD "9" Swallow Saloon - £250
WOLSELEY Hornet Swallow Super Sports 2-str. 12 h.p. 6-cyl. - - - - - £220
AUSTIN "7" Swallow Saloon - - £187-10
AUSTIN "7" Swallow 2-str. - - - - £165
Beautiful four-colour catalogue sent free on request.
SWALLOW
EXCLUSIVE COACHWORK
HENLYS
SWALLOW DISTRIBUTORS for SOUTHERN ENGLAND
Henly House, Euston Rd., N.W.1. Museum 7734 (30 lines).
Devonshire House, Piccadilly, W.1. Grosvenor 2271.
Manufacturers: The Swallow Coachbuilding Co., Coventry.
FACILITATE BUSINESS, and ensure prompt attention to your enquiries, by mentioning "The Light Car and Cyclecar" when writing to advertisers. They will appreciate it.

STANDARD
SWALLOW

4

April 24, 1931.
The Light Car & Cyclecar 23
SWALLOW......THE WORLD'S MOST BEAUTIFUL COACHWORK
Wolseley Hornet Swallow 2-seater Super Sports, 12 h.p., 6-cylinder model £220
Austin "7" Swallow Saloon £187-10
Standard "9" Swallow Saloon £250
A big claim this, but who will say that it is not amply justified? Where else can you find such a brilliant range of bodies as those illustrated here ... bodies of exquisite beauty and great strength ... finished in a selection of no less than twelve artistically blended colour schemes; coachwork which enables you to express your individuality in a pleasing and unique manner.
But mere words are inadequate. You must see the full range of models now on view at Henlys, or write to-day for illustrated catalogue.
Swift "10" Swallow Saloon .. £269
HENLYS
Sole Swallow Distributors for Southern England.
Devonshire House, Piccadilly, W.1.
Henly House, Euston Road, N.W.1.
Manufacturers: The Swallow Coachbuilding Co., Coventry
FACILITATE BUSINESS, and ensure prompt attention to your enquiries, by mentioning "The Light Car and Cyclecar" when writing to advertisers. They will appreciate it.

5

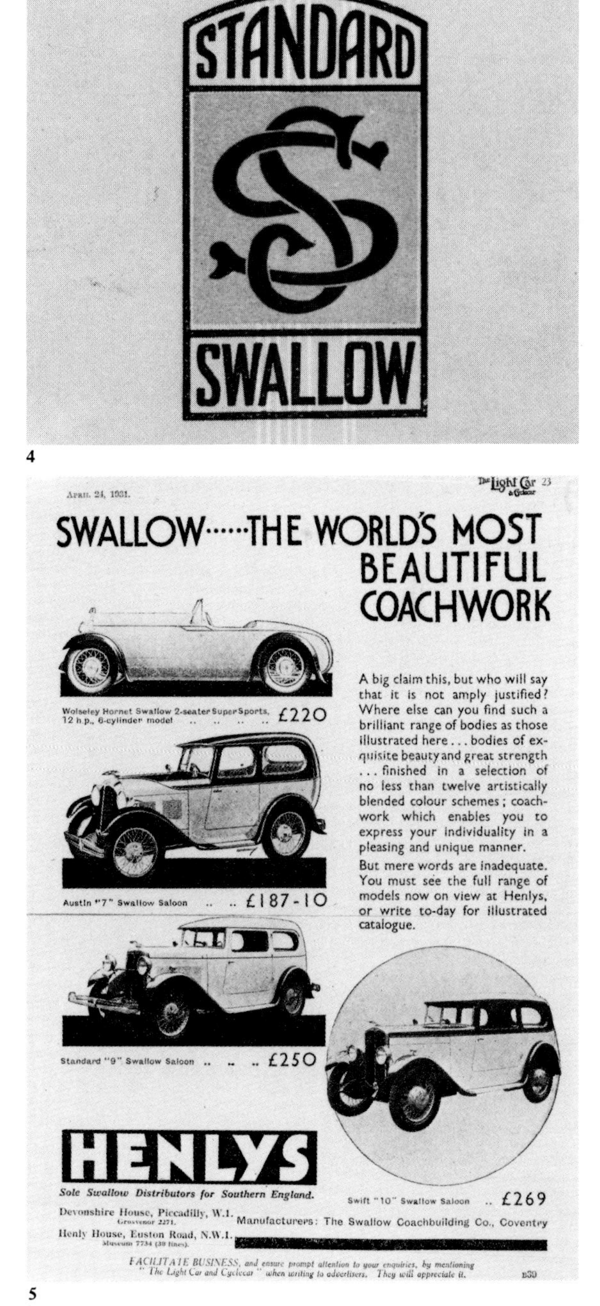

able front bucket seats, a rear blind with driver's control, Bakelite ignition and throttle-control levers, and a Yale-type door lock incorporated in the door handles.

In a period advertisement Swallow felt able to state that the revised car made 'an irresistible appeal to discriminating motorists', and the new features contributed to 'harmonizing to create one of the most outstanding small cars of the generation'.

This model continued to be very popular and the false dub irons were copied by other manufacturers. In January 1930, HH The Sultan of Perak took delivery of a Swallow Seven Saloon, appropriately finished in purple and black.

Visitors to the Motor Show of 1929 would have noted on the Swallow stand not only the familiar Austin but also no fewer than three new models. These were based on the Fiat Tipo 509A, the Standard Big Nine and the Swift Ten.

Walmsley had at first a Clyno and later an Alvis, bodied by the company for his own use. The company was always on the lookout for a new chassis for the Swallow treatment, especially after Lyons had been snubbed by Sir Herbert Austin. James Ritchie, later a Jaguar distributor, supplied the Fiat chassis, for which he was the Glasgow agent. Ritchie therefore approached Lyons suggesting that the Fiat might be suitable for the Swallow treatment.

The Fiat, like the others, was simply an enlarged version of the Austin Swallow theme. From the rear the two cars were very similar, but frontal treatment differed from model to model. The Fiat 9 hp chassis with an engine of 990 cc made a pleasant sporting saloon of modest performance. It had two doors and four seats, the rear ones being sunk below the chassis to give extra leg room. Production of this model ceased during 1930 after, it is thought, between 50 and 70 had been completed.

Most significant of the trio first seen at the 1929 Motor Show was the 9 hp Standard Swallow, for not only did it stay in production rather longer than the other two, with a consequently greater number being produced, but it began an association with the Standard Motor Company that was to assume greater and greater importance.

'The grace of line', claimed the brochure, 'appeals to the artistic taste.' Furthermore, the sales patter continued, 'The interior appointments signify the dignified refinements characteristic of the luxury car. The spacious lounge seats, heavily carpeted floors and polished mahogany furnishings enhance the superlative exterior finish.'

The most notable external difference from the other models was a new chromium-plated radiator shell slightly raked to the rear with a central bar coming down to meet the louvred tray between the dumb irons. The driver's windscreen 'opened if required when driving in fog or for ventilation', and a suction-operated screen-wiper was fitted to the lower part.

Considering its high standard of finish and range of appointments the Big Nine was not expensive at £245. 'In fact', claimed the company, 'there is no car, regardless of price, so fascinatingly attractive.' The following optional extras could be specified: Splintex glass for £10, a rear blind for 10/6d or front and rear double bumpers at five guineas.

A reader of one of the motoring journals wrote to the magazine in 1930 that she was the proud possessor of the first Standard Swallow to be put on the road. 'I gave my car a very careful running-in for well over 1,000 miles; then I set off complete with husband, friend, dog, portable wireless, three large and heavy suit-cases, also a superabundance of fishing tackle, to say nothing of three pairs of rubber boots (very bulky articles these) and a few other little odds and ends, for some river and lake fishing in mountainous Wales. The car behaved (as every lady should) perfectly, and revelled in showing the world that long and winding "one-in-sixes" were the joy of its life, also that it could jog along at a steady 45 mph all day and all night.

'Speed on top is about 57 mph but,

1

2

given good conditions, a mile or so more may be added. On second it is possible to attain 32 to 35 mph, but it is not recommended. While on the subject of gears, I should mention that to anyone not used to the Standard gear-change a little practice is necessary before it can be made quite silently.

'I find acceleration good, and in second gear very good. Petrol consumption works out at about 35 mpg at 25 mph. Easy starting. Good brakes. I have no grumble worth mentioning, and find my Standard Swallow a very reliable little car.'

The Swift Swallow was based on the Swift 'Fleetwing' chassis fitted with a 1190 cc engine. It had the usual Swallow refinements and many of the improvements introduced in the Swallow Austin in 1930, namely – leather upholstery, polished mahogany and double panel V windscreen. Two large doors gave good access to front and rear seats, the occupants of the latter benefiting from deep footwells. The usual adventurous colours were available plus Swallow Apple Green and Swallow Deep Leaf Green. Green, it was said, was a most hard-wearing colour, and it is interesting to note that the company paint shop mounted the drums of paint on electric motors, thereby agitating the liquid and maintaining consistency of shade.

The demise of the Swift concern in 1931 caused the cessation of this model after something like 150 had been produced at a price of £283 8s 0d each.

The relationship with Standard took a further step forward in May 1931 with the introduction of a Swallow-bodied 16 hp six-cylinder Enfield, which was priced at £275, with the optional sliding roof. In almost every way the car was identical to its smaller-engined brother. It is believed that 56 of these 'Ensign' Swallows were built, but the significance of the model was that it led to the company's acquaintance with the smooth, reliable 2054 cc sidevalve engine with its

1 *From March 1930 Swallow built a limited number of saloons based on the Swift Ten (J C)* **2** *The larger Standard served to introduce Swallow to the trusty Six, referred to in advertisements as a 15 hp* **3** *The most prominent feature of the Two-Seater Hornet was this pretty tail-end treatment, bravely flamboyant for the austere 1930s – seen here taking part in a Bugatti Club treasure hunt*

impressive 7-bearing crankshaft. To Lyons and Walmsley this sturdy engine seemed ideal for a project they had in mind.

Production at that stage was running at a good 100 cars per week, with the Austin Saloon still going strong. Indeed, the price was reduced from £187 10s 0d to just £165 for the saloon, and £150 for the two-seater with Coupé hood only, or £160 with the Coupé Saloon head only. Also, the list of optional extras was growing.

An ingenious little production line had been set up at the Coventry factory. The timber from the drying room passed through a special French machine which planed it on both sides, and thence into a circular saw. All wooden pieces were made with such precision that no alterations were necessary when they came to be assembled on jigs. As a result, eight lads could assemble more than 30 saloon frames a day. The panelling process took rather longer because of the need to eliminate every blemish that might be shown up by the high gloss cellulose which by then was standard.

A good chassis with lively performance successfully married to an exceptionally good-looking body, even by Lyons's standards, sums up the Swallow Hornet two-seater, launched in 1930. A number of contributors to the correspondence columns of the motoring press suggested that it was 'the most attractive small car on the road'.

The hood, quite shapely when erected, was completely concealed when folded, although it needed some practice to do this quickly and neatly. Two spare wheels, rather generous for a small car modestly priced at £225, were standard, and they were mounted on strong brackets on either side of the bonnet. The seat back rest which was not adjustable (to the chagrin of shorter drivers), hinged forward to reveal quite a large locker in the tail. The model sold well and continued in production until 1932.

A four-seater version of the Hornet

1

2

3

was listed from 1931. The rear end treatment was more traditional, with one rear-mounted spare wheel. The single pane windscreen was well raked and had triangular side panels. With comfort in mind, the newly designed front seats had cushions with a nine-inch front, which gave good support beneath the knees. The hood folded neatly and lay horizontally. Wings were close-fitting and had a high domed section for both style and strength.

In 1930 one of the magazines noted 'The Swallow Morris Minor – an experiment recently seen at Henly's Ltd, Euston Road', although this was purely a one-off.

In 1932 Wolsley announced the extensively modified 'Hornet Special', and Swallow continued with the two- and four-seater bodies to clothe this new variation.

The Wolsley company's philosophy was stated as follows: 'It is implied that the chassis is not sold as a racing job, but as a basis which is structurally correct and upon which keen coachbuilders can develop whatever they desire for the capability of high performance is present in the design; and all the much desired refinements for speed work are already incorporated.'

The bodies were identical to the previous models, 'the long lines (of the two-seater) emphasized by the narrow beading with the double lozenge beneath the windscreen'. A bare chassis would have cost you £180 and with Swallow two-seater coachwork a further £75, the four-seater being just £5 dearer.

1 *The Four-Seater clearly lost the style of the Two-Seater, with its average tail treatment* **2** *The Swallow Hornets continued to be produced concurrently with the early SSs* (JC) **3** *The frontal treatment meted out by Lyons to the Hornet Two-Seater – the Special was announced in 1932*

2 THE SS MARQUE

Just nine years from its inception the company evolved from mere 're-bodiers' to motor manufacturers. The SS marque achieved considerable publicity and created an affordable style that carried the company through the difficult thirties; and a stronger and most important symbiotic trading relationship was created with the Standard Motor Company.

1

2

3

4

William Lyons's ambition was to build his own cars rather than merely re-body other people's, for the latter necessarily places certain constraints upon your designs. To have a completely free hand you must produce your own chassis, and this he did in 1931 with the introduction of the SS I Coupé. The car caused quite a stir at its introduction at the 1931 Motor Show, because it had the look of a car costing £1,000 or more, yet it was listed at a mere £310. Lyons had once more brought style within easy reach of the motoring masses. A smaller model – the SS II – was also offered. The chassis and engine for both these cars were supplied by the Standard Motor Company, and it was said to be the first time that a chassis had been designed by a body-builder.

The initials SS have variously been ascribed to Standard Swallow, Standard Special, Swallow Sports, and Swallow Sidecars, but it was never really decided what it did stand for – it depended whether you worked for Standard or Swallow. The title may well have been borrowed from the famous Brough Superior SS 80 and SS 90 motor cycles.

A variety of body styles was very popular but, sadly, the performance of the car never really lived up to the expectations created by the sporty appearance.

1 *Formed in the early 1930s, the SS Car Club flourished until the outbreak of the war. This montage shows a good cross-section of models (*J C*)* **2–6** *'WAIT!', the advertisements had proclaimed in mid-1931, 'The SS is coming,' and on 9th October it duly arrived (*Mrs M. W. F. Longlands*)* **7** *A number of non-original additions (headlamps, horns, etc) make this a rather strange example of the very rare SS II Coupé, but it serves to show the overall shape (*National Motor Museum*)*

6

The car was unkindly parodied with slogans such as Soda Squirt and Super Sexed, but those ungracious people were going to have to eat their words in a few years' time.

'The SS is coming', announced the company's advertisement in the autumn of 1931. The new range, announced on 9th October, first appeared at the Olympia Motor Show later the same month. 'Long, low and rakishly sporting' was the press reaction. 'Boldly individualistic – daring to be different' claimed the SS catalogue.

By mounting the engine several inches farther back in the chassis than in the comparable Standard, and by lengthening the wheelbase and mounting the springs alongside the frame rather than under it, was the way in which Lyons was able to design such a low body for the period. This, plus the length of the bonnet, which was at least half as long as the whole car, contributed to its sporty appearance or producing, as one writer bluntly described it, 'a real cad's car'.

The frame was an entirely new design of the double-dropped type. From parallel dumb irons, the side members splayed out about two-thirds along the engine, whence they continued along parallel lines again. Level with the gearbox, the frame swept down $2\frac{1}{2}$ inches and then up again over the back axle. These side members were adequately cross-braced at various points. As already mentioned, the road springs were mounted alongside the chassis, and at the rear they were slung under the axle casing. This was a Standard component equipped with special hubs to allow the fitting of Rudge-Whitworth 'racing-type' wire wheels, shod with $5\frac{1}{2} \times 18$-in Dunlop tyres. The steering box, again of Standard manufacture, was fitted well forward to enable the column to lie almost horizontal, resulting in the desirable near vertical steering wheel.

All these modifications to the normal Standard chassis were designed in the

June 28, 1932. The Motor 31

The value of its beauty
...the beauty of its value!

S.S.1

No car within recent years has received such lavish praise as the S.S. "The car of the year" . . . "The car with the £1,000 look" . . . these are but two of the many expressions its beauty has evoked. In any company, in any circumstance, the S.S. looks and is distinguished.

And its value is no less striking than its beauty. For both the S.S. Chassis and Coachwork are made in the finest and best equipped factories of their types in the country. No other Speciality Coachbuilder has such an organisation and such facilities. And these are the principal reasons for the outstanding value of the S.S.

The S.S.1. Sports Coupe, as illustrated below, fitted with a 16 h.p. 6-cyl. engine. Ex Works, £310
The S.S.11. Sports Coupe, fitted with 9 h.p. 4-cyl. engine. In appearance, etc., a slightly smaller edition of the S.S.1. Ex Works, £210

See them at any of the many Swallow Agents throughout the country

A SWALLOW PRODUCTION

Manufacturers:
THE SWALLOW COACH-BUILDING CO., LTD., FOLESHILL, COVENTRY.
Telephone - - Coventry 8027

Distributors for Southern England:—
HENLYS, HENLY HOUSE, Euston Rd., London, N.W.1
and Devonshire House, Piccadilly, W.1.
Museum 7734 - - - (20 lines)

KINDLY MENTION "THE MOTOR" WHEN CORRESPONDING WITH ADVERTISERS. B27

1

2

3

4

interests of the ultra-low body. The result was successful. A man of average height could, while standing, comfortably rest his elbow on the roof, yet have adequate headroom when driving.

What the 16 hp Standard lacked in vivid performance it made up for in sturdiness, smoothness and torque. Steep gradients presented no problem, and the SS would accelerate without fuss from 5 mph in top. The low body obviously meant a low centre of gravity, which contributed to the above average roadholding. This was aided by the light, responsive steering and complemented by the easy SS-designed gear change mounted on the Standard 4-speed box. Both hand and foot brakes operated on all four wheels. Soon after its introduction, the SS I could also be had with the Standard 20 hp side-valve engine of 2552 cc.

The imposing radiator was V-shaped, slightly raked and surrounded by two large Lucas headlamps that were carried on a curved tie rod connecting the domed wings. The deep sides of the extravagantly long bonnet were covered with a plethora of louvres. Two large doors gave good access to the interior, which consisted of a wide bench just capable of carrying three adults, and a cramped rear seat adequate for luggage, two children or one adult at a pinch.

The rather heavy rear quarters of the roof were relieved by dummy 'pram

1 *The advent of the SS I once again gave the advertising copywriters full reign to exercise their verbosity (*J C*)* **2** *The SS I Saloon must have been considerably less claustrophobic for the rear seat passengers (*J C*)* **3** *An unusual rear view of the SS I Coupé showing the generously proportioned luggage locker* **4** *An SS I Coupé with one or two interesting extras, including the tinted sunshade (*Mrs M. W. F. Longlands*)* **5** *The SS I Tourer allowed the whole family to enjoy open-air motoring (*N. Baldwin*)* **6** *Proof that the SS I Tourer was indeed a full four-seater, with presumably a good deal of comfort (*J C*)* **7** *The SS I Coupé in its 1933 guise; with its flowing wing line it was a more balanced design*

6

irons', and to the rear was mounted a luggage locker and the spare wheel. A flush-fitting sliding roof was provided, and this and the rest of the roof were covered in black fabric. The car could be ordered in primrose with black wings and chassis or, if you preferred, lake and carnation red.

The SS II was simply a smaller version of the larger SS I and, as such, was inevitably overshadowed. That was rather a shame because, although it was a modest little car, it has come to be more highly respected with hindsight. Writing in the early 1970s a motoring scribe was moved to comment, 'The little SS is surprisingly nice to drive, bearing in mind that comfort is subordinate to styling . . . top speed is 60 mph if you're patient'. Based on a modified version of the Standard Little Nine chassis with a 9 hp engine of 1006 cc, it was priced at £210. It was purely a two-seater and although it lacked performance – 10 to 40 mph in top took 25 seconds – its fuel economy somewhat compensated for this with a consumption of 38 mpg and oil quoted at 2000 mpg.

The financial depression of the 1930s seemed to have little or no effect on SS sales, because the SS was probably the lowest-priced car available that gave the appearance of being expensively coachbuilt. And it was looks that still impressed many people in the 1930s.

To put into context the new SSs against their opposition, a 1932 price guide showed that the SS II at £210 was £5 dearer than the Standard Nine, £10 less than the Trojan and the comparable Renault, £18 less than the Rover Ten Special and £25 cheaper than the Hillman Aero Minx. The SS I 16 hp at £325, and the 20 hp at £335 compared with the Lanchester at £315, the six-cylinder Essex and four-cylinder Rhode at £315 apiece, the 16 hp Wolseley and the 19.9 hp Standard at the same price, the Chrysler Plymouth and Star at £345, and the Hillman Wizard, Morris Isis and Rover Speed Pilot all at £350.

For 1933 the SS I was revised to keep pace with fashion and improve on the previous model's shortcomings. The wheelbase and track were increased by 7 inches and 2 inches respectively, and the extra length was utilized in the rear seat area, allowing the previously cramped accommodation to be occupied by two adults, and therefore widening the market for this model. To enable this use of the back seats, the chassis was also modified, being underslung at the rear, passing under the axle and having triangulated cross braces. The rear seating area was treated in a novel way, because the propshaft split the back seats due to the low level of the body in relation to the axle, so the rear seat was sculptured like two armchairs. It is said, though unlikely to be true, that their design was based on the office chairs at Foleshill.

The individual helmet-type wings of the earlier model gave way to the continuous flowing type with running boards incorporated. And the tie-bar to which the headlamps had been attached was deleted, the lamps now being affixed directly onto the wings. Performance was increased a little by fitting aluminium heads to the trusty 16 hp and 20 hp engines, giving a higher compression ratio. Either a Solex or RAG-type carburettor and engine-driven fuel pump supplied petrol from the enlarged 12-gallon tank. A new design of radiator block and enlarged bonnet louvres with increased protrusion aided cooling.

On the road the SS I's strong point

was still its cornering, particularly at higher speeds. The long, flat, semi-elliptic springs gave a firm, yet not harsh, ride. A curved top panel to the opening single-pane windscreen improved vision, and the rear passengers, who had little or no vision, if suffering from claustrophobia could now at least have some ventilation from the opening rear window. Some relief could also be gained from new scuttle-mounted ventilators, while extensive use of felt around the bulkhead was intended to keep engine fumes out.

The upholstery, as before, was covered in Vaumol hide, and the front seats, now bucket-type, could be easily adjusted on their ball bearing slides. The tools that were formerly carried beneath the dash were now moved to the lid of the luggage trunk. Generally, the SS I had matured into a more balanced and practical car, yet with its distinctive individuality unmarred.

Another model was introduced in the spring of 1933, namely the SS I Tourer. This open four-seater had a pleasing appearance when in open guise, but the line suffered when the hood was raised. However, it gave a hint of what was to follow, and was the first model actively to campaign in competition. Although it

*The raised hood on the SS I Tourer did little to improve the Lyons' line (*N. Baldwin*)*

did not immediately win any major rewards, nevertheless the car did not disgrace itself. Henlys, the SS distributors for southern England, as they had been since Swallow days, waxed lyrical in their advertisements, stating, 'Just as the SS I Saloon has swept the country, soon this long, low, sleek open sports SS will be conspicuous on the roads everywhere'. At £325 for the 16 hp model and £335 for the 20 hp, the cars could not be said to be over costly.

Mechanically they were identical to the Coupés, and bodily they were very similar. Apart from the obvious alterations to make the cars open ones, the door tops were curved down for elbow resting and the scuttle top had a raised lip to deflect the slipstream when the folding windscreen was flat. The rear end sloped away with the spare wheel mounted at a rake where the luggage locker normally resided.

For the 1934 season yet another model was added to complement the range. This was a four-light saloon. A more genuine four-seater with the provision of rear side windows meant that the back seat passengers could at last see out. Because the saloon looked a more practical car, it appealed to the more conservative members of the motoring public. Mechanically it was identical to the Coupé and the Tourer. But that year the whole range underwent various changes. Engine sizes were increased to 2143 cc and 2663 cc, and the track was enlarged a further two inches to 4 ft 5 in. A stronger cruciform-braced frame allowed rear passengers more leg room, and the four-speed gearbox inherited synchromesh on second, third and fourth gears. Further improvements included self-cancelling trafficators (these had been an optional extra), Silentblock spring shackles, quick-action filler cap and reversing lamps, to say nothing of the chrome-plated band to the spare wheel cover. The most noticeable exterior change was the acquisition of a new and more imposing radiator without the previous bisecting slat. The hexagonal SS badge appeared on the exterior for the first time, mounted on this new revised radiator.

The SS II which, apart from the addition of a four-speed gearbox, had remained unchanged since its introduction, benefited in late 1933 from major revision and updating. Most importantly, the SS II was now provided with a specially designed low-built chassis along the lines of the SS I, with the wheelbase increased from 7 ft 5½ in to 8 ft 8 in, and the track by 1½ in to 3 ft 10½ in. This had the effect of making what was previously a very cramped little car into a small four-seater. The original cycle wings were discarded, to be replaced by the same flowing style of 'big sister'.

A Saloon supplemented the range. This was like the SS I, a Coupé restyled with rear side windows, and it made a very pleasant and popular small car. In 1935 these two styles were joined by a third when the SS II Tourer was announced.

When it came to engines for the SS IIs, you could choose either Standard's 10 hp or their 12 hp, and they would have cost you £260 and £265 respectively for the Coupé and Saloon, plus a further £5 if you chose the larger engine.

In late 1934, William Walmsley, who did not have Lyons's ambition, resigned from the company and involved himself in the manufacture of one of his hobbies, namely, caravans.

The company was by now producing a range of five models (soon to be increased), with an output of 1,800 cars a year from a factory site that had been expanded to some 13 acres. Once more it changed its name, this time to SS Cars Ltd, and was floated as a public company late in 1935, raising some £85,000. It is interesting to note that profits were approximately doubling each year, with £12,000 for 1931/2, £22,000 for 1932/3, and £37,000 for 1933/4.

Around this period more attention began to be given to the mechanical attributes necessary if SS were to make further progress. While the style of their products had achieved wide acclaim, they had suffered from jibes such as, 'more show than go', highlighting the fact that the cars' performance barely

1

2

lived up to the promise evoked by their sporting body designs. Just as Lyons had achieved independence of design by having his own chassis frames, so he wanted to move a step nearer complete independence regarding the power unit.

These desires led to two major personalities appearing on the SS stage. Firstly, Lyons consulted and contracted with Harry Weslake, who was building a reputation for his work with carburation and cylinder-head gasflow and design. Secondly, it was decided that the time had come for a proper engineering department to be set up. William Heynes, at 32, was appointed chief engineer, and he was to assume an importance in succeeding years second only to Lyons himself. After an apprenticeship with Humber, he was singled out and seconded to the design office where he soon found himself in charge of the Stress Office. When Lyons was looking for a suitable man to set up his engineering department, Grinham and Dawtry, who were in charge of design at Standard and had been Heynes's bosses at Humber, suggested the young man who had so impressed them. The task facing Heynes at SS was formidable. A new chassis had to be designed for the completely new model scheduled to be announced in five months' time.

In announcing the 1935 range, a further body style became available on the SS I chassis, this being an attractive Airline body. Company advertising proclaimed: 'The Airline Saloon – acclaimed from the moment of its introduction as the most beautiful interpretation of streamlining. This model is characterized by a modernity of outline, dignified in its restraint'. This style, in vogue at the time, featured a rounded tail incorporating a large boot, twin-mounted spare wheels in painted covers on either front wing, and very rakish horizontal bonnet louvres. Sadly, the model did not seem to catch on, which ironically accentuates its desirability today.

The year 1935 saw several mechanical improvements incorporated in the whole range. These consisted of a new high-compression head (Harry Weslake having advised here), bigger sump, higher lift camshaft and twin RAG carburet-

1 *An SS I Tourer supplied by P. J. Evans, one of the company's first agents (*Mrs J. B. Parker*)* **2** *The much revised and lengthened SS II, in saloon form, made the car rather less distinguishable from its larger stablemate* **3** *A representative of the Bolton constabulary receives their first SS II Tourer in 1935 (*J C*)*

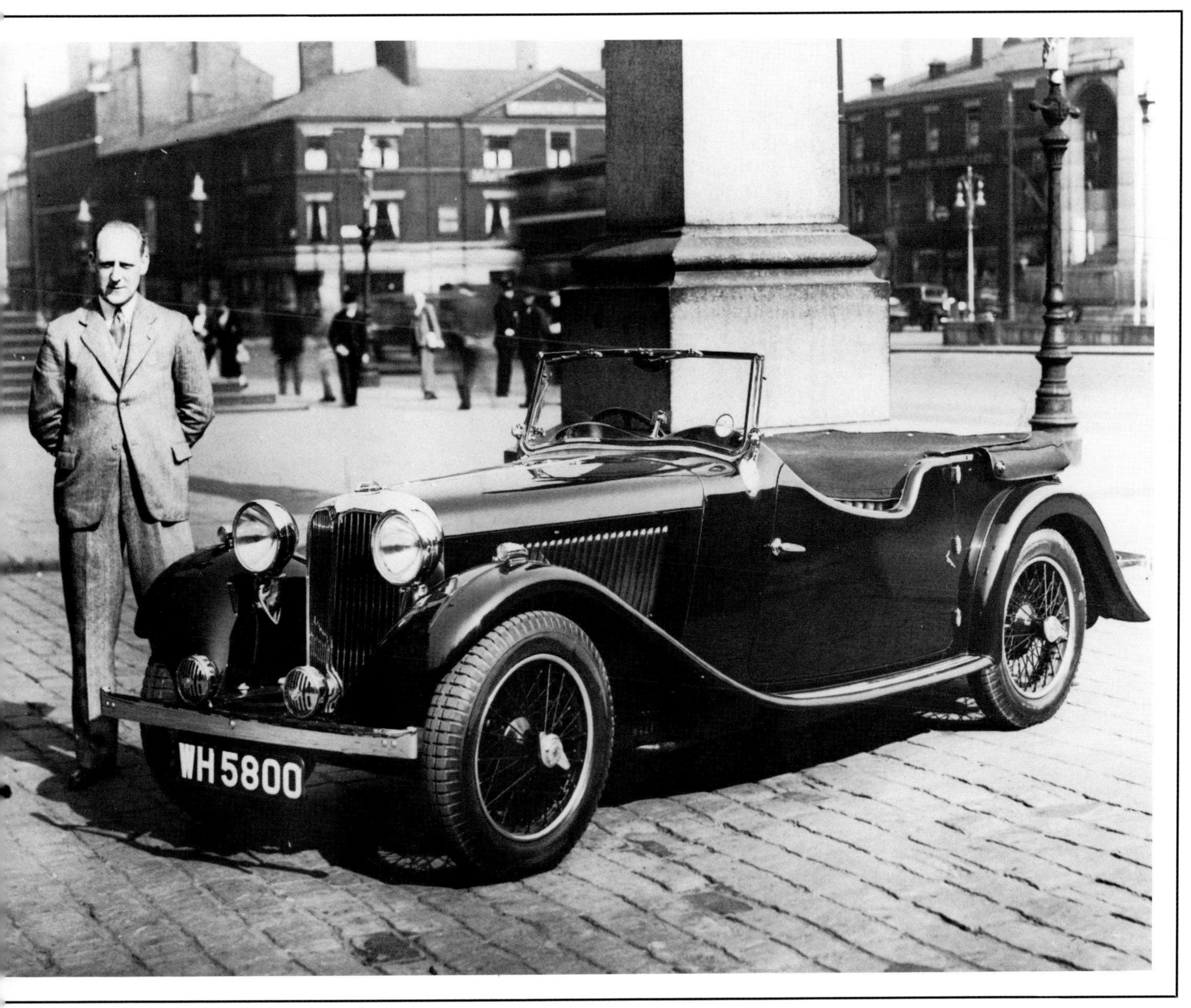

tors. The catalogue for that year listed metallic colour schemes for £5 extra. For the same amount, DWS four-wheel permanent jacks could be supplied, provided they were specified at the time of ordering. A period Philco car radio would cost rather more, at 16 guineas.

In March 1935, SS introduced the SS I Drophead Coupé. Mechanically identical to the Coupé, it also looked very similar but this time the 'pram irons' to the roof were no dummies but actually hinged the hood. This ingeniously folded down out of sight below a hinged cover, leaving only the closed 'pram irons' visible. To complete the effect, the window frames could be removed and stowed in the rear compartment.

Just 105 of these dropheads were built. They were priced at £380 for the 16 hp version or £5 more for the 20 hp car.

Also introduced in March 1933 was the new SS 90 sports car. This was fitted with the 2.7-litre side-valve engine and a close ratio gearbox. The prototype had a rounded tail with the spare wheel mounted into a recess. However, the production versions had what was, for the period, the almost obligatory slab tank rear end, with the spare wheel mounted vertically. The basis was an SS I chassis shortened by some 1 ft 3 in and fitted with André Telecontrol shock absorbers. As the name implies, the top

1

2

3

speed was reputed to be 90 mph, and for that privilege you would have to pay a modest £395. The SS 90, of which only 23 including the prototype were built, was quickly overshadowed by the SS 100, due to be announced in only a few months' time.

Lyons took an SS 90 to an SS Car Club weekend event at Blackpool and, although not officially competing, demonstrated the new car over a speed test course on the sea front to record a time that was nearly seven seconds quicker than the fastest official competitor. His pleasure at this must have been somewhat lessened when he heard that the secretary of the club had absconded with the club funds, leaving Lyons to settle the not inconsiderable hotel bill.

1 *With the adoption of a new Swallow-designed chassis, the enlarged SS II Coupé became a full four-seater (*Mrs Bayne*)* **2** *The SS Airline was certainly a departure from normal Lyons thinking, and the last time he was to allow fashion to dictate to him* **3, 4** *The SS I Drophead Coupé : the final manifestation of the SS I line and surely the happiest, with its beautiful folding 'wig top'. (*National Motor Museum; J C*)* **5** *The SS 90 : a beautifully blended design but as yet without the power to match the expectation suggested by the style (*J C*)*

3 THE SS JAGUAR

The name Jaguar made its first appearance on new models that were more sober in style, more akin to the Bentleys and similar cars of the period, but gained much in practicability. Competitive pricing ensured value-for-money and as technical standards and methods of production improved, prosperity continued. Only the outbreak of war could temporarily arrest the company's progress.

1

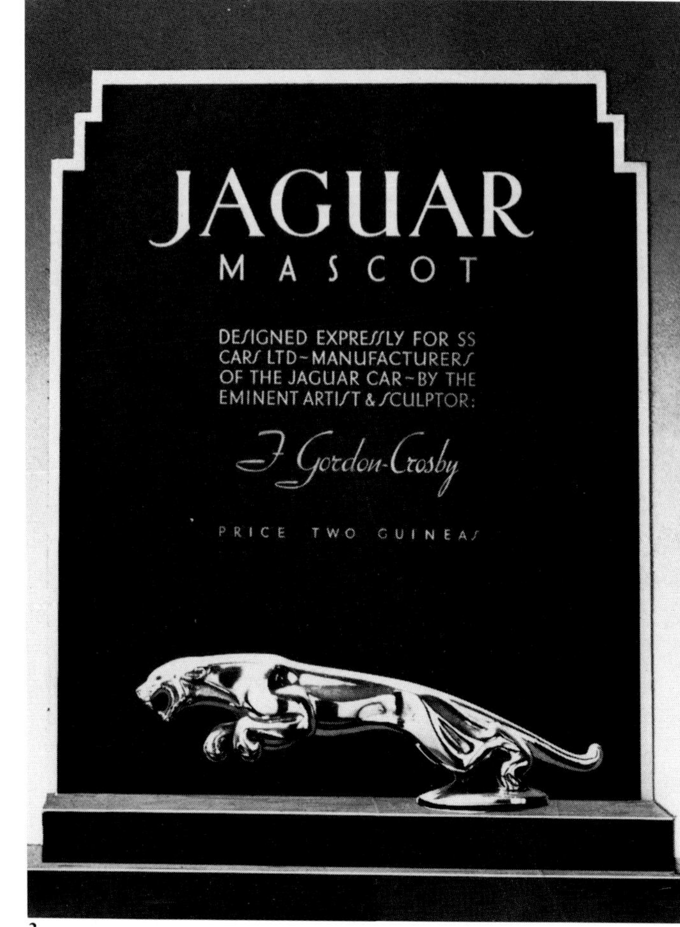

2

3

4

The year 1935 brought a new range and a new name, for these new models were to be known as SS Jaguars. At first intended only as a model name to be applied to the two new saloons and the SS 100 sports car announced in September of the same year, the name of Jaguar acquired greater significance as time passed. This significance was so great that one of the major English dictionaries defines the word as 'jaguar (jagew-er) n. large yellowish spotted carnivorous animal of cat family; (tr) (cap) make of powerful motorcars'. It appears to be the only make of car to be accorded this compliment because the only other car names to be listed, such as rover, morris and triumph have merely their traditional definitions noted.

The new name was chosen after the publicity department had prepared a list of birds, fishes and mammals for Lyons. He had no hesitation in choosing Jaguar with its allusions to excitement and speed, and the memories it held for him of stories he had been told about the Jaguar aero engine.

The publicity department was soon at work glorifying the new models. 'The car of the future has arrived. Swift as the

1 *The 1936 SS badge, incorporating the new model name of Jaguar (*J C*)* **2** *An early advertisement for the famous mascot that has come to be known as 'the leaping Jaguar' (*J C*)* **3** *An SS Jaguar 2½ ohv saloon at rest with Eastnor Castle in the background (*National Motor Museum*)* **4** *A period shot of the 2½ saloon showing its touring capabilities (*Mrs E. Nicholson*)* **5** *The splendidly proportioned lines of the 2½ ohv tourer, photographed in California*

wind – silent as a shadow come the SS Jaguars', claimed a lavish six-page pull-out advertising feature in the *Autocar* of 18th October 1935. 'Effortless speed combined with perfect Town "manners", make driving a sheer delight'.

The new SS Jaguar saloons were endowed with a completely new Lyons-designed body. For the first time, the company offered four doors and this, plus the better proportioned shape, made for a less dramatic appearance but more practical and comfortable accommodation. It had the usual Lyons style, and indeed the shape was a step nearer to the contemporary Bentley, which cost nearly four times as much. Lyons, with his flair for the dramatic, arranged a special lunch a little before the model's announcement at the 1935 Motor Show. Those present were then asked to estimate the price of the new car. The average guess was £632. The actual price as listed was a mere £395.

Weslake had contracted to extract a minimum 95 hp from the 2½-litre, which yielded 75 hp at that time. His plan was for an overhead valve configuration, so he designed a new cylinder head with pushrod-operated overhead valves and reshaped ports and combustion chambers. The results to everyone's satisfaction, was an output not of 95 hp but of 105 hp. Furthermore, Lyons managed to persuade Standards to acquire the machinery to produce these new heads, thereby saving SS the vast capital outlay necessary had they been forced to produce them themselves. The completely new Heynes-designed box section, cruciform-braced chassis frame was no longer of Standard manufacture but was made by Rubery Owen.

Lyons had now succeeded in producing a practical saloon with style and performance, and all at a price that his competition could barely approach, let alone match.

Announced concurrently with the saloon was a tourer body, which was also fitted with the 2½-litre OHV engine in the new chassis. The combination of the lively new power unit and the stylish traditional SS open four-seater bodywork must have made for a very pleasant motor. However, not a great many were made in the two years that they were listed, and they may have been overshadowed by the new two-seater sports car.

The tourer was very similar in external appearance to the earlier SS I Tourer, with the exception of the new chromium-plated radiator grille of the saloons. And if you looked very closely you might also notice the larger 15-inch Girling brake drums and, of course, the Jaguar badge.

About that time Bill Rankin, who was head of publicity, designed the famous leaping Jaguar mascot which, with some minor revision by the famous artist F. Gordon Crosby, has become closely associated with the marque. Contrary to popular opinion, this was an optional extra, and not fitted as standard until 1957.

Alongside the 2½-litre models there was a 1½-litre version. Similar in external appearance to the larger-engined version, this model retained the old side-valve engine, which meant that these cars were sadly lacking in performance compared with say, the MG of the period. However, at just £285 they offered remarkable value for money as ever, and were very well appreciated. Both the new SS Jaguar saloons had side-mounted spare wheels, the 1½-litre being recognizable, apart from its shorter front wings and bonnet, by the top of the spare wheel casing protruding just above the bonnet line.

The 1937 brochure stated, a little hypocritically, 'Self-praise is no recom-

1

mendation', and proceeded to print the following:

Capt Sir Malcolm Campbell
169 Piccadilly
London W1

Dear Sirs,
I was so impressed with the styling qualities of the new SS Jaguar that I decided to buy one myself and you will be interested to hear that I have in consequence placed an order with your London agents, Messrs Henlys Ltd
Wishing your company continued success,
Yours very truly,
M. Campbell

Pre-empted by just six months by the SS 90, the SS Jaguar 100 (to give the car its full name, although today it is more usually known as the SS 100) was announced with the saloons and tourer in September 1935. Visually very similar to the SS 90, the 100 can be identified by its slightly slanting fuel tank and spare wheel, whereas these were dead vertical on the earlier model. More minor differences included the headlamp tie-bar, which had 100 cast into it instead of 90, different headlamps, and subtle changes to the radiator shell which had, as in all models, the winged SS emblem above the name *Jaguar*.

The chassis followed earlier SS lines but with later suspension and steering, and had a wheelbase of 8 ft 8 in. The

1 *Considered by some at the time to be too flamboyant, the SS 100 is now thought by many to be the epitome of prewar sports car design (*J C*)* **2** *The spare wheel clearly protrudes above the bonnet line, distinguishing the $1\frac{1}{2}$ sv saloon (*National Motor Museum*)* **3** *The most famous product of the company's prewar era: the SS Jaguar 100 (*Mrs E. Simms*)* **4** *One hundred miles an hour indeed! What will these young 'uns be up to next? (*Mrs E. Simms*)*

4

price for the SS 100 two-seater 2663 cc sports car was just £395, for which you had a car capable of 95 mph with a 0–60 mph time of around 13 seconds. A company advertisement claimed that the model, 'though primarily for competition work, is sufficiently tractable for use as a fast tourer without modification'.

For many, the 100 was the climax of Lyons's prewar designs. It was beautifully proportioned with gracefully flowing lines that imbued the car with a sporty feel. It also gained genuine sporting success, attracting attention in Europe – testimony to the fact that the performance was beginning to match the promise.

A large number of small changes were made to the saloons and tourer in 1937. Externally the 2½-litre acquired quarterlights, P100 headlamps and faired-in sidelights. The tourer was fitted with P100s and the 1½ had the quarterlights. All models featured a reorganized interior, resulting in more comfort with greater leg room.

Late in 1937 there was further progress when a new range of models was announced to replace the 1½-litre sidevalve saloon and 2½-litre ohv saloon and tourer. The new range, offered in 1½, 2½, and 3½-litre forms with either a saloon body or drophead coupé style, were now of all-steel construction. The former method of construction, whereby steel panels were attached to a wooden frame, was superseded, and the use of wood was dropped. This new method allowed production to be increased, with the result that the total output approached 100 cars per week during 1938. But this transition to all-steel construction was not achieved without considerable trauma. Thousands of pressed panels were bought in from several suppliers, but when SS attempted to assemble them they just did not marry up. As a result, the factory stood virtually idle for several months while the situation was resolved, and so serious was the situation that bankruptcy became a possibility. Happily, everything was eventually sorted out and the new methods allowed production to be increased dramatically. This resulted in a certain trading loss being transformed into a respectable profit.

Heynes designed a new chassis for the all-steel range. Keeping abreast of the then current thinking, this consisted of 6-inch-deep side members of box section with three box-section cross members. The greater stiffness achieved obviously aided the ride of the new range. All three saloons shared the same body, which closely resembled the previous 2½ with some subtle changes. The spare wheel disappeared from the wing, and was attached at first to the boot lid and later to a tray under the boot floor. The new bodies were somewhat larger with improved interior space and better access through larger doors.

The 1½ now had a wheelbase of 9 ft 4½ in, as opposed to the larger-engined cars with a wheelbase of 10 ft. This difference was achieved by a correspondingly reduced bonnet and 'cut-and-shut' chassis for the smallest version.

The 2½-litre engine remained substantially unchanged apart from improvements to its breathing arrangements, with the provision of a two-branch exhaust manifold. The 3½-litre was offered with a new engine. It was in principle just a larger version of the 2½-litre, still basically Standard but with the Weslake-designed head and other refining by Heynes. With a bore and stroke of 82 × 110 mm, Twin SU car-

1

burettors and two triple-branch exhaust manifolds, this new unit gave around 125 bhp at 4500 rpm.

Country Life, in reviewing the new model, stated that 'the $3\frac{1}{2}$-litre Jaguar is a very remarkable car indeed and may be honestly ranked among the high-class high-performance cars of the world. . . . If one had not looked at the catalogue before trying it out, one would be forgiven for thinking it was at least in the £1,000 class.'

All three models were also offered in the attractive drophead coupé form. A well designed hood meant that the shape was pleasing whether the hood was raised or lowered.

The price of £298 was modest for the $1\frac{1}{2}$-litre, but then so was the performance, with such a large, heavy body and relatively small engine. However, the luxurious appointments perhaps compensated for the lack of zest. In contrast, the $3\frac{1}{2}$ litre, with its power output of 125 bhp, ensured lively performance, and a price of £465 for the drophead coupé must have made this a most attractive model in more senses than one.

1 *This shot shows the flowing Lyons lines, but here with an unusual positioning of the number plate and rear lamp (J C)* **2, 3** *The SS 100's hood was obviously a necessity rather than a styling feature (J C), whilst its dashboard was a delight to the eye (Mrs E. Simms)*

It was only logical to fit the new big engine to the 100, especially as it now had a greater chance of attaining its sporting aspirations. The $3\frac{1}{2}$'s healthy output, combined with the lightness of the sports body, gave a very lively performance indeed for the period. The 'ton' was just possible, and 60 mph could be reached in a shade under $10\frac{1}{2}$ seconds.

At a price of £445 there was nothing to touch the 100 for its combination of performance and price, to say nothing of its simple good looks. In other departments the SS 100 was also rated highly. The brakes were particularly good, and the roadholding inspired confidence. But the car did not escape criticism entirely. Tom Wisdom commented that 'the main fault of the SS 100, even in $2\frac{1}{2}$-litre form, was that it had too much power for its rather flexible chassis. The springing, in the sports car fashion of its day, was as light as all Hell, so instead of absorbing bumps it just ricochetted the car from one bump to the next. At the front anything over half an inch of suspension movement brought the bump stops into play'. Yet, he went on to say that in spite of the magazine road testers' disbelieving the performance figures at first (such was the acceleration), nevertheless the SS was 'a brilliantly docile and flexible car'.

With all the new development work it is not surprising that Heynes was hard-pressed and began to search for an

1

2

experienced man to assist him. Writing in 1970, Tom Wisdom stated, 'At my suggestion "Wally" Hassan, famous for his work on various Thomson and Taylor specials, including the record-breaking Hassan Special, and also on the ERA project in its formative stages, was wooed away from Thomson and Taylor at Brooklands and onto the SS payroll. Here his outstanding qualities as an engineer found good scope in the development of the bigger, 3½-litre SS 100 which came onto the market in late 1937.' Hassan joined the company in late 1938 with the title of Chief Experimental Engineer.

Apart from the current models, another model, most striking in appearance, graced stand No 126 at Earls Court in 1938. Lyons had designed a closed coupé body to clothe a normal 100 chassis and the result was an interesting hint of things to come. The show car had the 3½-litre engine and was priced at £595. Additionally, a 2½-litre was listed at £545 but was never produced, because only the one coupé was ever built. It had a four-

1 *Models for 1937 at the 1936 Olympia Show, distinguishable by the addition of quarterlights* (N. Baldwin) **2** *The nicely proportioned 1½ litre saloon brought luxury to the small car class* (J C) **3** *The 3½ saloon – a 'Bentley' at a quarter of the price – costing £445* (J C) **4** *The drop-head models were not entirely 'all steel', retaining some timber framing in their construction* (National Motor Museum)

speed synchromesh gearbox giving ratios of 3–8, 4–58, 7–06 and 12–04 to 1, and the car weighed 25 cwt.

Consideration was given to comfort and there was said to be more elbow room than one might suppose. There was a large space for luggage at the back, with the rear parcel shelf opening to a horizontal position to increase carrying capacity. This same flap contained the tools in a 'well-designed tray'. The rear wings were faired inwards at the rear to reduce 'windage', and the spare wheel was carried in the tail. Side lamps were faired into the front wings, and Lucas P100s completed the front lighting equipment.

The subsequent owner who received the car as a 17th birthday present found it not altogether practical. Apparently, some modifications had to be carried out to the interior to make it drivable, the doors dropped, the small sunshine roof created a whirlwind effect, and the red colouring on the steering wheel rubbed off on the driver's trousers! But, in spite of the practical problems, the model's significance in the evolution of the company's designs cannot be overstressed.

As stated earlier, sidecar production, although inevitably overshadowed by the more glamorous cars, nevertheless continued apace. With the floating of SS Cars Ltd, Lyons had formed a separate company, Swallow Coachbuilding Co (1935) Ltd, with a capital of £10,000, to carry on the manufacture of sidecars. The 1938 catalogue listed no fewer than 12 models plus the Special Swallow Chassis at £8 15s 0d (less tyre and tube).

Leafing through the catalogue, we find the Syston Sports at £17 19s 6d; the Sports Touring Coupé de Luxe with quick-lift hood operated by passenger from inside; the Hurlingham Adult

1

Two-seater 'evolved as a direct result of the numerous enquiries we have received for a two-seater sidecar'; the Kenilworth Coupé 'which will make a direct appeal to the family man who is desirous of accommodating both passengers under cover'; the De Luxe Touring Model with fashionable Vauxhall-like fluting and 'constructed on the pressed steel principle'; the Tourer De Luxe, as above but with 'a dickey-seat at the rear capable of carrying a normal-sized youth up to fourteen years of age'; the Standard Launch, which looked as though it should be on water rather than on land; and the De Luxe Launch, which with its added deck rails looked like a streamlined coffin. Still turning the pages, we find the Aero Launch Coupé, the coffin with a hard-top; the Sun Saloon, a fastback coffin; the very rakish Donington Special for sporting events; and finally, at £31 10s 0d, the Ranelagh Saloon, 'the finest sidecar obtainable'.

After a couple of years of 'all-steel' production, World War II intervened and car production was phased out, although the public was destined to see the saloons and drophead coupés again after an interval of some six years.

1 *By folding back the cant rails you could easily assume the Coupe de Ville guise of the three-position hood (*National Motor Museum*)* **2** *The very beautiful if not entirely practical one-off SS 100 Coupé, as shown at the 1938 Motor Show (*The Motor*)* **3** *The frontal aspect, although cluttered by later badges and horns, gives a hint of things to come (*J C*)* **4** *In parallel with the saloons, the sidecar range by the late 1930s showed more subtlety of line (*J C*)*

1

2

3

4

4 THE WAR AND AFTER

Aircraft repair and sub-contract work kept the factories busy during the war and the new machinery introduced and techniques learnt were to be of future benefit. The early post-war period was a dull time devoid of new models but it was also a period of consolidation with exciting projects at an embryonic stage. The company was becoming increasingly more independent, particularly of Standards, and much more of a complete manufacturer.

1

Towards the end of 1939, war work was commenced and car production gradually faded out. In comparison with the rest of Coventry, the Jaguar factory escaped serious bomb damage, although in late 1940 six workshops were destroyed. Apart from some experimental work, the company increased production of the old faithful sidecars for military use, supplying them to the RAF, the Admiralty, the Army and the National Fire Service; carried out a variety of aircraft production and repair work; and produced some 700 trailers of various kinds per week.

The aircraft work included the repair of Whitley bombers and later, of Wellingtons; the manufacture of various tooling; the production of wings, frames and bomb doors for Short Stirling bombers, Avro Lancasters, Spitfires, and de Havilland Mosquitos; and parts for the Cheetah seven-cylinder radial engine. Towards the end of the war the company was instructed to manufacture the complete centre sections for the then highly secret Gloster Meteor III jet fighter.

Other work included lightweight trailers to be towed behind motorcycles and jeeps, a folding sidecar, mule carts for the Burma campaign, and two lightweight jeep-type vehicles. The last-named were designated the VA and VB, and were intended for parachute dropping. The VA had a V-twin JAP engine

1 *Armstrong Whitworth Whitley bombers: some 80 sets of wings were modified by SS cars to carry heavier bombs (J C)* **2** *Ironically the war took SS back to where it all began, with the major emphasis placed on sidecars instead of cars (J C)* **3** *SS fitters working on the reduction gear of an Armstrong Whitworth Whitley bomber (J C)* **4** *The disciplines of war were much in evidence here (J C)*

2

3

4

but, because this was not considered powerful enough, the VB was built with a Ford 10 unit. What is particularly interesting about these vehicles is that they were of unitary construction – a method used for Jaguar's production cars some years later, and they were independently sprung all round, again preceding the production cars by approximately 20 years. But these vehicles never proceeded beyond the prototype stage, because by the time they were sufficiently developed, aircraft were capable of carrying the larger, more conventional jeeps.

The war had both bad and good effects on the company's progress. Lyons, having made record profits of £60,000 in 1938/9, had purchased one of SS's panel suppliers – Motor Panels – with a view to becoming independent of outside sources in this major sphere. But because it was starved of profits, the company did not have the money to recommence car production and reluctantly sold Motor Panels to Rubery Owen in order to finance postwar production.

However, one important benefit of the war work was the breadth of experience gained and the extra machinery acquired. The experience of aircraft manufacture and technique was to manifest itself later, particularly in the field of competition.

With hostilities drawing to a close, thoughts returned to car production. New models were planned, but with steel shortages, rationing, and other problems, it was thought best to re-introduce the prewar models as a stop gap.

At an extraordinary general meeting

1

2

3

in March 1945, it was decided to change the company name because the initials SS had by that time acquired an unhappy connotation with the Nazis' use of this title. So the company acquired the name of Jaguar Cars. This was a name that was to become increasingly familiar across the Atlantic, because Lyons had realized the value of that potential market and was beginning to exploit it. At home, however, with austerity at its worst, you had to wait anything up to eight years for a new Jaguar.

The sidecar business was sold to the Helliwell Group, who in turn sold it to Tube Investments. This company produced, alongside the sidecar range, the Swallow Gadabout – a 125 cc scooter – and the Swallow Doretti sports car, based on the Triumph TR2. In 1956 TI sold the Swallow Coachbuilding Co (1935) Ltd to Watsonian of Birmingham.

1 *Close on 100 Gloster Meteor III centre sections were built in the later stages of the war (J C)* **2** *VA jeep: SS began to apply the construction methods learnt from aircraft to the lightweight jeeps for parachute dropping, developed for the War Department (J C)* **3** *VB jeep: a variation on the theme but with more power, provided by a Ford 10 side-valve engine – note the interesting tyre pattern (J C)* **4** *Immediate postwar restrictions dictated the re-introduction of the prewar range, but postwar austerity did not preclude traditional opulence. The thinner chrome band, seen here on a $1\frac{1}{2}$-litre, was one of the few features that distinguished it from the prewar versions (J C)* **5** *The Jaguar window transfer shows that this $2\frac{1}{2}$-litre is a postwar model (J C)*

Shortly before the end of the war, Jaguar's took another step forward by acquiring from the Standard Motor Company all the machinery used in the manufacture of the 2½- and 3½-litre engines. This ensured the company's independence and, with hindsight, was viewed with regret by Captain Black of Standard's.

This early postwar period was a difficult one for every firm – not least for Jaguar. It was an era of strikes, power cuts, steel shortages, coal shortages, and so on. As stated, the prewar models were re-introduced in mid-1945, in almost identical form. The 1½-litre was supplemented by a 1½-litre special equipment model, which had been available just prior to the war. It had a heater, better headlamps, adjustable front seats, foglamps and a 'specially finished luggage locker'. The only visible distinguishing features between pre- and post-war cars were a rather thinner waist band on the latter models, plus different badges and the word *Jaguar* replacing the SS on the wheel spinners.

The 2½-litre and 3½-litre models were also re-introduced in mid-1945. The only under-the-skin changes from their prewar counterparts were as follows. A Metalistik crankshaft torsional vibrational damper was incorporated, as were Girling 2LS brakes, and a hypoid final drive for both models. The revised final drive was also fitted to the 1½-litre.

1

2

3

At that time, with exports assuming immense importance (particularly to the American market), the $3\frac{1}{2}$-litre was especially popular. The Americans were used to big engines and fuel consumption bothered them little; whereas in Britain, with petrol rationing in operation, the thirst of the $3\frac{1}{2}$ was unacceptable, and the $1\frac{1}{2}$, despite its lack of performance, was preferred for its thrifty consumption.

Company advertising claimed that 'every Jaguar is a full five-seater car of high performance, luxuriously appointed with that impeccable finish which for years has been associated with the name Jaguar'. They now felt able to call it 'The Finest Car of its Class in the World'.

1 *A gentleman of the constabulary appears to be having some difficulty in entering this $1\frac{1}{2}$-litre* (J C) **2** *The $3\frac{1}{2}$-litre Jaguar, a rare sight in car-starved Britain in the late 1940s when the home market took second place to export* (Mrs E. Nicholson) **3** *Proving that nothing is new, a team of ex-servicemen, having formed a rural travelling maintenance team, service a market gardener's $3\frac{1}{2}$-litre in Worcestershire* **4** *The $1\frac{1}{2}$, $2\frac{1}{2}$ and $3\frac{1}{2}$ litre models were produced as a drophead and most were exported.* **5** *Superb styling, but not for claustrophobic rear seat passengers* (National Motor Museum) **6** *The Mark V – successful stopgap between prewar continuation and postwar innovation* (J C)

6

The drophead coupés were again offered from December 1947, and the vast majority of these were exported. Another important overseas market for Jaguar products, perhaps second only to the United States, was Belgium.

The Belgian Jaguar distributor, a Madame Bourgeois of Brussels – an ebullient character by all accounts – was also an energetic and fervent propagator of the Jaguar cause. Sales flourished – so much so that the Belgian Government banned the importing of all cars costing more than £600. Lyons replied by arranging for his cars to be assembled at the Vanden Plas works in Belgium. This had the added result of making the Jaguars more competitive in price. The Belgians dropped the ban in 1949, the factory closed and the production of mainly drophead models ceased.

In September 1948 the company announced its first new postwar model, although it was actually another stopgap compromise. This model was named the Mark V, for no better reason than that the production car was the fifth prototype to be built.

Standard's had stopped production of the $1\frac{1}{2}$-litre engine so the new Mark V was available in just $2\frac{1}{2}$ and $3\frac{1}{2}$-litre form. A new chassis was introduced which, according to the company's brochure (a lavish production with individually affixed colour illustrations rather like giant cigarette cards), was 'scientifically designed and, for its weight is probably the most rigid frame incorporated in any passenger car'. The stiffness of this frame contributed to the Mark V's good roadholding and ride. The adoption of 16-inch rather than 18-inch diameter wheels reduced the centre of gravity and improved the car's appearance. A newly developed 6.7 section Dunlop tyre

1

2

3

improved grip and comfort.

Most significantly, the car featured independent front wheel suspension, which had been originally designed in 1938 and continually tested and developed since that date. It worked on the torsion bar principle first developed by Citröen and long favoured by Heynes, who gradually developed it into a very neat system.

The new Mark V could be had in drophead coupé form although this animal is extremely scarce in Britain because the vast majority, like the earlier models (sometimes unofficially referred to as Mk IVs), were exported, mainly across the Atlantic.

Like its predecessors, the new car had a three-position hood – fully raised, coupe-de-ville (open over the front seats only) and fully open. In any of these positions the model looked elegant and the car-starved British public must have been extremely envious of those in other countries who were given precedence at that time. The Mark Vs were tough, reliable souls if not terribly exciting, but excitement aplenty was just around the corner.

1 *The Mark V continued Jaguar's association with the police, as an envious public waited for its new Jaguar* (J C) **2** *Two-toning suited the Mark V, the first production Jaguar to carry spot lamps* **3** *The Mark V's interior – British tradition at its best—oozing luxury* (J C) **4** *The contrasting curious coachwork in the background illustrates the traditional line of this export Mark V* (National Motor Museum) **5** *The Mark V Drophead, still stylish but a little less happy with the spat, which contributed to the general heaviness of the rear* (J C)

5

5 THE XK ENGINE

The introduction of the famous new engine was to be the most significant step forward yet and helped guide the fortunes of the company right into the 1980s. In a period of austerity, when 'export or die' was the dictum, the new Jaguar sports cars and saloons brought a sense of pride to the British, doing much for international prestige, reinforced by patriotic competitive success.

1

As far back as 1939, a new engine had been planned by Lyons and chief engineer Heynes, but the war intervened and, although they discussed ideas during fire-watching at the factory, the actual design was not laid down by Heynes and his team, Wally Hassan and Claude Bailey, until 1946. They were given an entirely free hand in the design, striving to produce an engine with a minimum output of 160 bhp, extreme smoothness and flexibility, a capacity for continuous development and, interestingly, Lyons specified that the engine must be a 'glamorous' one in the manner of the Bugatti.

That they succeeded, and almost to the letter, is now a well known fact. The resulting engine, named XK, is arguably the greatest and certainly the most versatile ever produced. It proceeded to power exciting production sports cars, Le Mans winners, a variety of saloons in the 1950s and 1960s, Scorpion tanks and armoured cars, Dennis fire engines, record-breaking power boats and, in the 1970s and early 1980s, silent, sophisticated executive saloons.

The XK120 was designed and built as a publicity exercise and as a mobile testbed for this new XK engine. The intention was to build a limited number, prior to the engine's being used in an important new large saloon car to be introduced a little later. This was the Mark VII, duly announced in 1950. But the acclaim and orders for the XK120 were such that the car quickly came to be thought of as a serious production

1 *HKV 455 – a very well known number plate seen on the first XK120 Open Two-Seater Super Sports. Note the straight screen pillars, which were fitted to very early cars. (*National Motor Museum*)* **2** *The XK120's impact was immeasurable, attracting attention from far and wide, typified by these distinguished visitors (*J C*)*

2

venture. More than 10,000 were built, most of which were exported to America.

Although Heynes, Hassan and Bailey had discussed ideas for a new engine during the war, they started in earnest to design it only in 1946. The first engine design was a four-cylinder job designated XA, X standing for experimental. However the first engine to be produced was the XF, a four-cylinder engine of 1360 cc (66.5 × 98 mm). Its crankshaft was not sufficiently strong, so next came the XG. At that time Bill Heynes was very friendly with the well known sports racing driver Leslie Johnson. Johnson had a BMW 328 which Heynes borrowed and converted one of the then current Jaguar pushrod engines to the same type of head. This XG had a capacity of 1½ litres. The engine was installed in a BMW and Walter Hassan 'ran this car for thousands of miles', testing not only the engine but also various types of suspension, including the 'Lockhead air type'. But this engine was too noisy and was superseded by the XJ, on which basic specification the production engine was to be based. This XJ design was produced as a 2-litre, four-cylinder and a 3.2-litre, six-cylinder. The latter was found to be lacking in low speed torque, so the stroke was increased, giving the final capacity of 3442 cc. These engines, after exhaustive testing (mainly on the four-cylinder) and minor adjustments and alterations, were given the designation XK. The cylinder head with hemispherical combustion chambers and cross-flow design had been the work of Harry Weslake. On an 8:1 compression ratio the larger engine produced 160 bhp at 5,200 rpm, and it also looked right, with its polished aluminium camshaft covers. Maximum bmep was 140 lbs sq in at 2,500 rpm, and bhp per square inch of piston area produced a figure of 3,175, and peak piston speed in feet per minute was 3,360.

In the summer of 1948, before the public knew of the car that was to come, the works had lent an XJ four-cylinder engine to the well known record-breaker, Major Goldie Gardner. He used it to good effect in his streamlined car EX 135, breaking three International Class E records with a mean speed of 176.694 mph – from just 2 litres.

The situation in 1948 was that Jaguar had no sports car currently listed, yet they had a new chassis and a brand new high performance engine. So with the 1948 Earls Court Motor Show not very far off, the decision was taken to build a sports car. There were several reasons for this. Lyons realized from the SS 100 the value of publicity resulting from successful racing competition, resulting in a sporting image being acquired for the Jaguar company as a whole. But additionally, it was an ideal opportunity to try out and, hopefully, prove the new engine in a low-production model prior to the large-scale production envisaged when the Mark VII was announced. Furthermore, the sports car was likely to be bought by enthusiasts, who would be more tolerant of any teething problems.

The decision having been taken, it was now up to William Lyons to speedily design a body. The result won almost universal acclaim and a measure of the rightness of this example of the Lyons line is the high esteem in which the shape is held today, more than 30 years later. You must remember that in 1948 an all-enveloping body was virtually unheard of for a sports car, the only precedents being the BMW and a few one-offs designed by Italian specialist coach-builders. And certainly the standard of

1

2

comfort and protection offered was unrivalled.

In almost every way the XK set new standards – in performance, styling, ride, roadholding, docility (for such a car), comfort and, above all, in value for money. The factory was able to claim that it was the Fastest Production Car in the World.

The chassis and suspension evolved for the Mark VII, used in the interim Mark V and slightly modified for the sports car, consisted of two deep box members running from the front to just in front of the rear axle, where these members gradually decreased in size and swept up over the axle. The front suspension consisted of independent wide-based top and bottom wishbones and long torsion bars located around the middle of the chassis. Rear springing was by semi-elliptics and lever-type shock absorbers, and Lockheed full hydraulic brakes worked in 12-inch

1 *In 1948 there was nothing on the road that could keep up with an XK120 (*National Motor Museum*)* **2** *An early steel-bodied XK120, complete with its 'all weather' equipment (*J C*)* **3** *Not surprisingly the exciting XK120 led the great Jaguar postwar export boom, many being ordered from the USA (*National Motor Museum*)* **4** *The frontal aspect of the wings makes an interesting comparison with the prewar 100 Coupé – the spotlamps were an extra (*B. Bradnack*)* **5** *Occupying the XK120's cockpit must have been every schoolboy's dream in the early 1950s*

5

drums. The gearbox was a four-speed one with synchromesh on second, third and fourth, and the ratios were 3.64, 4.98, and 12.29 to 1, giving approximate speeds of 62 mph in second and 90 mph in third. The standard axle ratio was 3.54:1 which translated, allowing for changes in tyre radius, to 4951 rpm at 120 mph in top. Dunlop Road Speed tyres were fitted with normal pressure of 25 psi all round, but 'for fast driving when comfort is not of primary importance' they could be pumped up to 35 all round. Lucas de Luxe electrical equipment was fitted throughout with 'twin batteries with constant voltage-controlled ventilated dynamo'.

For £998 (£1298 with purchase tax) you had a car that could better 120 mph, accelerate to 60 mph in 10 seconds, or potter gently through rush-hour London. So remarkable were the factory's claims that many of the public and indeed most journalists found that they could not believe them, dismissing them as exaggerated publicity.

Jaguar had a problem. The public and press alike needed proof in a form that was conclusive. So, a party of journalists was flown to Jabbeke motorway in Belgium, where a carriageway had been closed. The Belgian RAC had set up official timing apparatus and an XK120 with R.M.V. (Soapy) Sutton was ready and waiting.

Lofty England recently recalled, 'Soapy was a very nervous character until he got going. He used to stand there shaking, smoking madly and dropping ash everywhere, particularly down himself! Anyway the car was got ready including tying the sidescreens to the hoodsticks so they wouldn't fly off, and the plane was late. There was ash everywhere!'

The car was a standard example with the addition of an undershield, an optional extra. First run, with hood and side screens in position, achieved 126.448 mph. Then the windscreen was detached, a small cowl replaced it and a metal tonneau was attached to cover the passenger seat. This run produced a speed of 132.916 mph in a northerly direction and 133.596 mph in a southerly direction – a mean speed of 132.596 mph. To conclude the

1

demonstration, Sutton motored past the amazed journalists at 10 mph – in top gear.

Despite their incredulity, the public had not been slow in placing orders for the new car, and it was realized after a few days of the Show that demand would considerably outstrip the 'couple of hundred' originally envisaged. The first cars were constructed in aluminium over ash frames. Obviously this method was not suited to mass production, so it was necessary to tool up for pressed steel bodies. This inevitably meant long delays, and although 240 aluminium-bodied cars were made, it was not until mid-1950 that the first steel-bodied cars were produced and that XK120s became available in any quantity. Even then the majority went abroad, mainly to the States and to Australia, the company's largest export markets at that time.

In 1950 William Lyons, not content solely with being President of the Society of Motor Manufacturers, again stole the show at the 35th International Motor Exhibition at Earls Court. The star on this occasion was the first entirely new Jaguar saloon to be presented since the war, the culmination of five years' research and development – the Mark VII.

As has already been mentioned, the twin-cam engine was designed and developed not for a sports car range, which was merely a convenient mobile test-bed, but for a completely new big Jaguar saloon. This model was the Mark VII, and the XK engine gave this very large (overall length 16 ft $4\frac{1}{2}$ in) and heavy (33 cwt dry) car a top speed of 100 mph and, for the period and type of car, lively acceleration, with a 0–60 mph time of $12\frac{1}{2}$ seconds.

Apart from the engine, the saloon and the sports car had much in common

1 *A solid-wheeled car, this XK120 is correctly fitted with spats; the separate chromed front sidelights were fitted until 1953* **2** *Works driver 'Soapy' Sutton preparing for his famous 132 mph run at Jabbeke* (J C) **3** *In 1953 Jaguar showed this XK120 MC complete with 18-carat gold-plated wire wheels (note later sidelights and front wing ventilators)* (National Motor Museum) **4** *The Mark VII – the start of the 'Grace, Space and Pace' era* (J C)

3

4

mechanically. The chassis designs were similar, being a straight plane steel box section frame of immense strength. Suspension and steering followed XK120 lines closely. The brakes, a Girling Autostatic fully hydraulic self-adjusting system, not surprisingly because of the weight of the car, had servo assistance. The standard of ride and smoothness was impressive and led Jaguar to claim that the car gave 'the impression that the engine is merely idling when the car is travelling at high cruising speeds'. After all, the engine was only revolving at a shade more than 5,000 rpm at the ton, and a piston speed of 2,500 ft min produced a top gear speed of 69 mph.

The interior, with its vast leather-covered seats (it was described as five-seater, six optional) and an almost excessive amount of walnut, positively oozed with luxury, redolent of prewar opulence. Publicity material was at pains to point out that 'four large suitcases and four big golf bags can be carried, with room still left for sundry small items of hand luggage'. The boot had a capacity of 17 cubic feet, and large cubby lockers were provided on either side of the instrument panel, one fitted with a lock and key.

Not surprisingly, the Americans immediately coveted the new saloons which had, after all, been designed with that market in mind. Within two or three months, orders had been taken to the value of approximately 30 million dollars, and Lyons had a real big volume winner. This inevitably imposed production problems and necessitated a move from Holbrook Lane to the present Browns Lane site where one million sq ft of space allowed considerable expansion and a resultant rise in output.

1

2

3

It is interesting to note that Lyons, who had tried to imitate the Rolls Royce and particularly the Bentley in appearance, was now approaching them on technical specifications as well. This narrowing of the gap has continued to the present day. Incidentally, there was no Mark VI Jaguar because Bentley were already using this designation, and Jaguar's adoption of the title Mark VII precluded Bentleys from continuing their numerical progression – no doubt a source of some annoyance to them.

The XK120 Fixed-Head Coupé was introduced at the Geneva Motor Show in 1951. It was basically a roadster with a roof but to put it just like that simply does not do justice to the marriage, which was another Lyons success.

Reminiscent of prewar Bugattis, the FHC was altogether a more comfortably appointed car than its stablemate. It had wind-up windows, figured walnut fascia and door cappings, veneered instrument panel and better heating and ventilation. The performance was little different from that of the roadster, although a special equipment model was offered with $\frac{3}{4}$-inch lift camshafts, a dual exhaust system, and 8:1 compression, developing 180 bhp. These SE models also featured wire wheels, a special crankshaft damper, and stiffer 1-inch torsion

1 *In spite of its size, the new XK engine propelled the Mark VII at 100 mph (*J C*)* **2** *Interior of a Mark VII belonging to Major Goldie Gardner, the famous record-breaker, competing in a Coronation Concours d'Elegance at Brighton (*National Motor Museum*)* **3** *A Jaguar Mark VII helps to perform the opening ceremony of the new Motor Industries Research Association (MIRA) test track (*N. Dewis*)* **4** *Part of a glamorous early 1950s filmstar image: poodles and Jaguar Mark VII (*National Motor Museum*)*

bars. They were designated XK120 Ms in the States and were available in roadster form as well. Acceleration from 0–60 mph now took some 9.9 secs, and the fixed-head weighed in at some 27 cwt as opposed to $25\frac{1}{2}$ cwt for the original alloy cars, and $26\frac{1}{2}$ cwt for the steel roadsters. A couple of features first seen on the closed 120s were also adopted for the roadsters, namely footwell ventilators and sidelights faired into the front wings.

The fixed-head was a Grand Touring car in the proper manner, and in right-hand drive form is a rare car today because only 195 were produced. It was stated in the factory-produced Salesman's Data Book for 1954 that the 'Special Equipment model is produced in series and is, therefore, eligible for acceptance in competitive events for production sports cars'.

Stirling Moss had an XK120 FHC with which he used to tow a caravan to meetings. When I asked him recently if he remembered it, he replied. 'I remember it very well because the caravan broke loose and smashed! But I do remember the 120. It was a very pretty car and in its time handled quite well. It was a very good long-legged car. Suited me well.'

The design and construction of the C-Type is covered in the competition section, because the C-Type was designed purely with competition in mind – hence its original title of XK120C, the C standing for competition.

First raced in earnest in June 1951, the factory offered a 'production' version in August of the following year. This was to allow private owners, particularly across the Atlantic, to compete against the sports racing Ferraris, etc, on equal terms.

The C makes an extremely good road car as I can testify, having had on one occasion to try and keep up with an extremely rapidly driven D-Type along country roads. Yet the car is docile

1

2

1 *Jaguar show stand,* circa *1953, featuring the Le Mans winning C-Type, XK120 Roadster, XK120 DHC and Mark VII (*J C*)* **2** *Not so much a racing machine like its open sister, the 120 FHC admirably suited the man with sporting aspirations at the weekend (*National Motor Museum*)* **3** *The C-Type made a fine road machine, and this example was obviously used as such, judging by the tax disc and registration number (*National Motor Museum*)* **4** *With its added sophistication and comfort, the XK120 Fixed-Head made a good, fast, long-distance touring car (*J C*)* **5** *The C-Type belonged to the era when competing cars could be and were driven to events such as Le Mans (*N. Dewis*)*

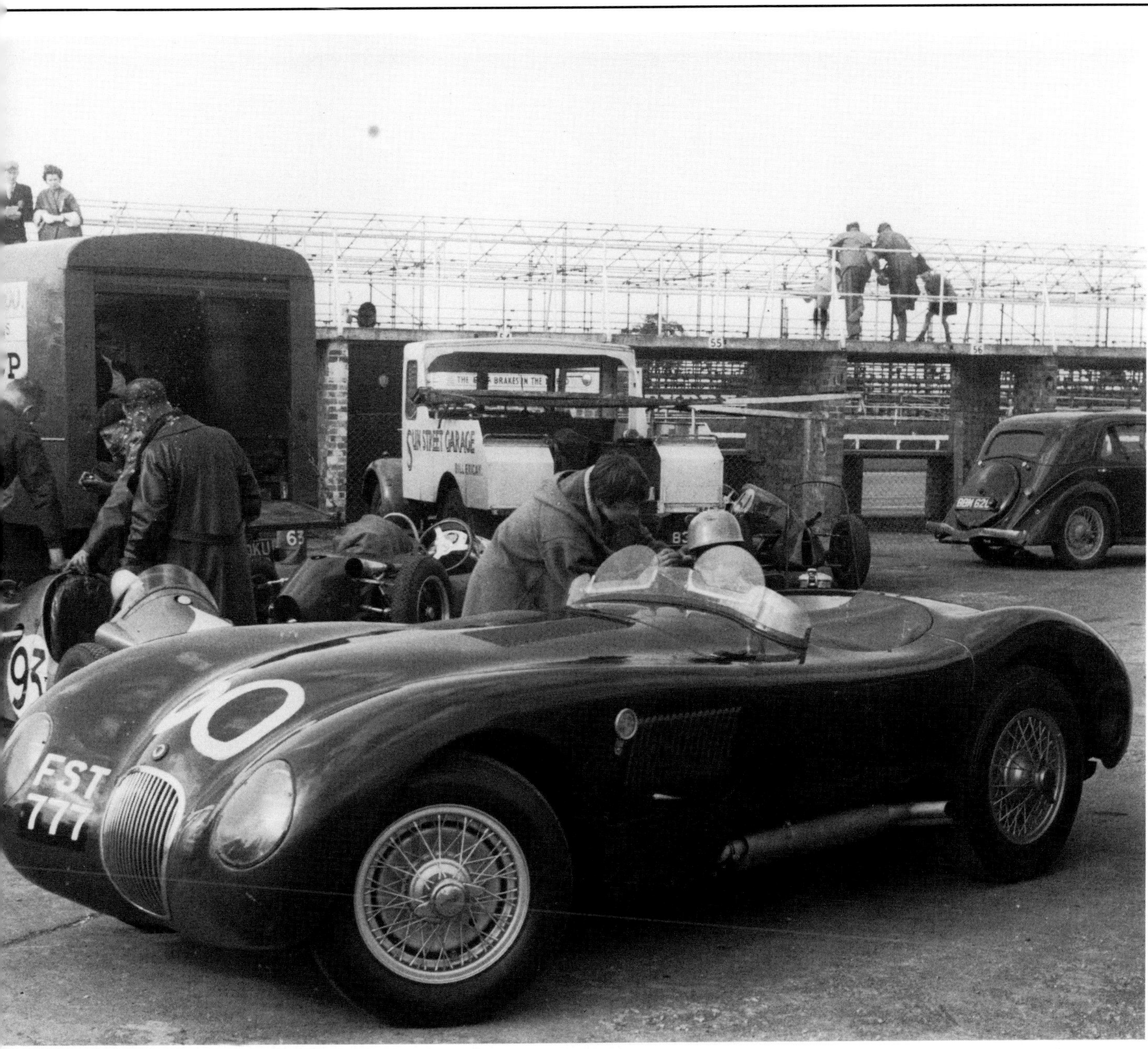

5

Announced at the Geneva Show in 1951 and seen here at Earls Court the same year, the breathtakingly beautiful Fixed-Head XK120

CARBODIES
JAGUAR

1

2

3

4

enough to trundle through towns, although there is not much space for the shopping.

Including the works cars, 53 C-Types in all were produced. The production version cost just £2,327, including purchase tax of £832. There were no fundamental differences between these cars and the one which had won Le Mans in 1951. With road use in mind, some concession was made to creature comforts with basic trimming of the cockpit. Also, a horn was required by law, a law that permitted the 140 mph plus of which the C-Type was capable. There was no other car in production (even though production was limited) at that time capable of keeping up with a well-driven C-Type.

Of the 53 C-Types that were built, all but about half-a-dozen were production Cs, but irrespective of whether they were works cars or not, they were numbered in the sequence XKC 001 to XKC 053.

Introduced in April 1953, the XK120 Drophead Coupé was a compromise between the out-and-out sporting Roadster and the more sophisticated fixed-head. Its mechanics were the same as those of the other models, but it was fitted with a lined, folding 'first quality padded mohair' hood, with an interior light and a rear window that could be unzipped. Interior trim was similar to that of the more refined fixed-head. The result was a pleasant and practical car that looked attractive with the hood up. But it was a little spoilt by the bulky folded hood, which left rather a pronounced bulge to the line, just behind the doors. This model is the rarest of all the XKs, because only 1,765 were produced, compared with 2,678 fixed-heads and 7,612 Roadsters.

The colours available for the XK range of coachwork at that time were as follows: Suede Green, Cream, Birch Grey, Battleship Grey, Lavender Grey, Black, Pastel Green, Pastel Blue, Dove Grey, British Racing Green and Red.

1 *The XK120 Drophead Coupé announced in 1953 completed the trio, with most of the fixed-head's comforts but still with the option of open-air motoring* (J C) **2** *Apart from the C-Type, the entire range for 1953 was on display in the Birmingham showrooms of P. J. Evans, still today one of Jaguar's major distributors* (R. Gore) **3** *Concours events featuring car and owner were a popular pastime in the 1950s and here is an unusual two-toned XK120 DHC* (National Motor Museum) **4** *A mouth-watering display of Jaguar's products, including a lone C-Type, awaiting despatch. All but a handful are left-hand drive* (J C)

With no less than four greys listed they sound almost dull compared with the Swallow colour schemes.

It is interesting to note in this day of increased awareness of fuel consumption, that an XK120 competing in the 1954 Cheltenham Motor Club Economy Run achieved the remarkable figure of 58.7 mpg.

At the beginning of 1953 the Americans were courted yet again with the introduction for the Mark VIIs of an optional automatic transmission that would not be available on the home market for a couple of years. This was followed in 1954 by the further option of Laycock de Normanville electrically selected overdrive on the manual models.

During the previous year Norman Dewis, Jaguar's chief tester, had taken several cars over to Jabbeke in Belgium, including a Mark VII, with which he managed to record a mean speed of 121.7 mph.

At the 1954 Motor Show, coinciding with the introduction of the XK140s with the uprated 190 bhp engine, was the revised Mark VIIM, which utilized the same engine together with stiffer suspension and closer gear ratios. Externally only minor differences were evident, with Le Mans type headlamps, foglamps moved to the extremities of the bumpers and replaced by horn grilles, altered bumpers without the middle shallow rib, and the previous semaphore-type indicators giving way to the later flashing type. To combat body roll the torsion bars were increased in diameter from $\frac{15}{16}$ in to 1 in. Inside, the pointed horn push first seen in the XK120 was changed to the flat, and safer, XK140 type.

The D-Type, the most famous of

1

2

3

Jaguar's sports racing cars and arguably the most famous postwar sports racing car of all, was made available on a limited production basis from the middle of 1955.

The construction and specification of the D-Type is covered in the competition section, but suffice it to say here that by this time the works D-Types had already had a fair amount of success, including a win at Le Mans earlier in 1955. As a result, there was a demand from private owners both here and in the States who wished to use them either in competition events or purely as very exciting road cars.

An actual production line was set up and some 42 cars were completed (not including 9 which were either destroyed in the factory fire or dismantled, and 16 which were made into XKSSs).

Although they were built on a production line, rather more care in construc-

1 *The Mark VIIM, recognizable externally by its valence-mounted spot lamps, horn grilles and indicator flashers* **2** *Probably the most exciting, exhilarating seat of any car produced in the 1950s, the D-Type combined fantastic performance with complete docility (*J C*)* **3** *The Production D-Types allowed private owners, particularly in the USA, to enjoy a fair degree of success in sports car racing (*J C*)* **4** *The XK140, the 120's successor, with the much heavier bumpers and cast grille much in evidence, but otherwise with unchanged classic shape, here in Roadster form (*J C*)* **5** *The XK140 Roadster's interior was basically unaltered, apart from detail changes such as a better steering wheel rake, a flatter horn push and greater leg room (*J C*)* **6** *This XK140 Roadster perhaps benefited from the removal of its front bumper, often the most criticized feature of the new range (*P. Porter*)*

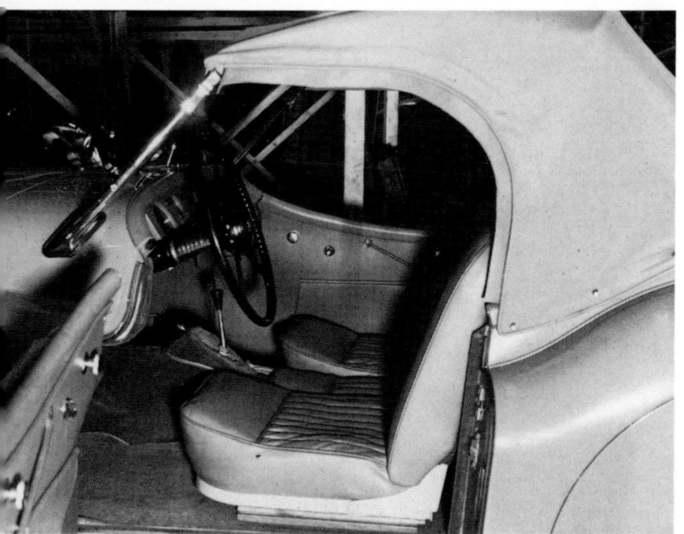

6

tion and detail work was taken, to say nothing of the testing (mainly at MIRA) prior to delivery. This was exhaustive, with one car covering as much as 650 miles on test. Of these 42 production Ds not surprisingly the largest number (some 18) were exported to the States, while the remainder were delivered as follows: Britain 10; Australia 3; France 2; and Cuba, Finland, Spain, Belgium, Mexico, New Zealand, East Africa, Canada and El Salvador received one each.

Although most of the more serious private owners received works cars, nevertheless the production D-Types have between them gained over the years a tremendous number of successes up to, and including, the present. The last one to be built, in the hands of Equipe Nationale Belge, finished fourth at Le Mans in 1956 and 1957.

Apart from competition activities, the production D-Types have, over the years, provided a relatively select number of owners with sensational road performance even by today's standards, for an original price of £3878 (I can testify to this, having been lucky enough to sample Nigel Dawes' example recently).

The XK140 introduced at the Motor Show in October 1954 was something of an enigma, because although in retrospect the car did not have the glamour of its predecessor, and the pure line was rather upset by its over-heavy bumpers, it was mechanically a superior car.

The special equipment engine of the XK120, which now produced 190 bhp, was fitted as standard, while the XK140 was also available with the C-Type head (introduced in April 1953 to the 120s) when the power rating was 210 bhp. For the first time on an XK, overdrive was offered as an optional extra, and this was found to be a very worthwhile feature, contributing to the XK140's image of a 'grand touring' car rather than a pure performance sports car. Cooling was improved with a more efficient radiator, raked to avoid the new steering rack. Contemporary road tests differed widely in their findings regarding performance, but it generally seems that, in spite of increased weight, the overall performance was slightly better than that of the XK120.

Right from the XK140's introduction all three body styles were offered. The Open or, as it has come to be known, the Roadster, differed least from its predecessor visually. Internally little was altered, although the two six-volt batteries were no longer carried behind the seats but were replaced by one twelve-volt situated in the near-side front wing. Externally, the new models were distinguished by the larger, rather over-heavy bumpers, one piece at the front and quarters at the rear. In similar vein a heavy cast grille replaced the 120's more delicate item, better headlamps and separate wing-mounted flashing indicators were incorporated, and chrome strips were added to the bonnet and boot lid, which was now several inches shorter and did not carry the number plate as previously.

It was in roadholding that the most significant changes came about. Firstly, the engine was moved three inches farther forward in the chassis. This had the effect of improving cornering by making the car more controllable with a weigh ratio of 50.3% on the front rather tha the previous 47.5%. This created mor leg room in the cockpit – which was slight deficiency in the XK120 Secondly, the Burman recirculating ba type of steering of the former car wa replaced by an Alford and Alder rack and-pinion. This latest change brough complimentary comments from drivers who felt that the steering was lighter an more precise, and the turning circle wa increased from 31 ft to 33 ft. More mino changes involved the fitting of uprate torsion bars and Girling telescopi rather than lever-type shock absorber at the rear. UJs incorporated in the steering column allowed the steering whee

1

to be set at a more appropriate angle.

Some 3,354 Roadsters were produced in all, of which only 73 were right-hand drive, and only 47 of those were intended for the British market, making this a rare model indeed. Total price, including purchase tax of £471, was £1,598.

All the new XK models shared the same mechanical specifications and slightly revised external alterations described for the Roadster. However, the drophead had another revision over the similar model of the 120, notably the provision of two occasional rear seats. This created what would be described today as a two-plus-two. These two seats might be adequate to carry two extra adults down to your 'local' providing it was very local, otherwise they were really only suitable for children, but at least it allowed the family man to enjoy XK motoring a little longer.

One can only surmise where the idea for the extra seats came from but the following may give some clues. A letter in the 28th September 1951 edition of *Autosport* from a Mr. B. W. J. Hindes of Slough read as follows: 'I thought you might be interested in the attached snap which shows my own car fitted with a third seat, which enables three fully grown adults to sit in complete comfort with the hood erected. This would surely increase the "saleability" of the vehicle (after all, we haven't all got aggressive mothers-in-law!) and I wonder why the manufacturers do not incorporate this feature, which in my case was carried out by a local coachbuilding firm with very little trouble.'

Another possible inspiration may have come from the competition world. Late in 1953 Ian Appleyard had a chance of clinching the European Rally Cham-

1 *The Americans inevitably took the lion's share of XK140 Roadster production. This example is fitted with wind deflectors, although they were not a factory option (*National Motor Museum*)* **2** *Announced in 1954 concurrently with the Roadster and Fixed-Head, the XK140 Drophead Coupé was positively civilized with its folding hood and walnut dash and extra seats (*National Motor Museum)

pionship if he entered and was successful in the Norwegian Viking Rally. However, the regulations for this event called for four seats and therefore his 120's (RUB 120, the replacement for NUB 120 – see competition section) roadster body was replaced by a drophead body to which modifications were made to enable two small seats to be fitted behind the main ones. In fact, RUB 120 was not used on this rally because it was thought to be un-British to try and beat the rules, but it may have been the inspiration for the 140 dropheads, of which 479 rhd and 2,310 lhd cars were built at just £46 more than the open two-seater.

The 140 DHC was the all-rounder XK, open but with a civilized hood with larger rear window than the 120, reasonable interior space, more or less the classic line and mechanically superior to the 120s.

Of the trio, the fixed-head was the most altered. Sharing all the modifications of the Roadster and the drophead, excluding the adoption of a single battery but including the occasional rear seats, the FHC went a stage further. In addition to the extra 3 inches gained from moving the engine forward, the bulkhead was modified to surround the engine, thus giving greater leg room and allowing the seats to be moved forward some 12 inches. This, plus extending the back of the roof by a further $6\frac{1}{2}$ inches, combined with the moving forward of the windscreen a fraction, provided substantially more room. A larger rear

1

2

window and larger rear quarter-lights contributed to this effect.

The fixed-heads were fitted with plunger-type door handles as opposed to the then more common lever type, as can still be seen on the dropheads. All the 140s at 14 ft 8 in were 3 inches longer than their predecessors because of the new bumpers.

List price for the fixed-heads was £1,616 for the basic model, and a further £214 for the Special Equipment 210 bhp model. Some 2,808 fixed-heads were produced in all, made up of 843 rhd and 1,965 lhd models. In the States, as before, the Special Equipment models were known as 140Ms, the M standing for 'modified'. When fitted with the C-Type head they were designated XK140MCs.

1 *In the XK140 Drophead Coupé only minimal space was allocated to the so-called extra seats (JC)* **2** *The 140 DHC had a pleasing shape even with the hood raised, but although fitted with a larger rear window than its predecessor, rear threequarter vision was restricted (*J C*)* **3** *Clearly visible is the extra length of the 'fixed-head' portion of the XK140 FHC (*J C*)* **4** *Looking happier with wire wheels, the 140 did not shirk an occasional competitive event (*National Motor Museum*)*

6 THE XK EVOLVES

Whilst the Jaguar sports cars brought the 'Pace', and the large saloons the 'Grace and Space', the company needed a production car to boost turnover and the new compact range filled that niche in the company's armoury. The later fifties saw gradual improvement in all models rather than dramatic developments – but these were at the planning stage.

1

Three years of design and development work and one million pounds' worth of tooling yielded the new 2.4-litre small saloon in 1955. This was a most important model which, with refinements and a variety of engine sizes, was to be in production until 1969. It started to appear on the roads in 1956 – a remarkable year for Jaguar and Sir William Lyons, because he was knighted in the New Year's Honours' List. In the same year that Jaguar won at Le Mans and Reims, they won the famous Monte Carlo Rally, making the company the first ever to win both Le Mans and the 'Monte', and a further compliment was accorded the company with an official visit by Queen Elizabeth II and the Duke of Edinburgh. The remarkable XK engine continued to power these winning cars, the compact range, the large saloon and, from 1957, the ultimate development of the XK range – the XK150s.

But all was not rosy during this period because on 12th February 1957 fire broke out in the factory and spread rapidly. It was brought under control before it reached catastrophic proportions but enough damage had been done to upset production considerably. The damage was thought to total £3½ million, but with ingenuity and co-operation the factory and production recovered

1, 2 *In 1956 Sir William Lyons and Jaguar were accorded the honour of a Royal visit, as Her Majesty inspects the Le Mans winning D-Type with her host and visits the shop floor (J C)* **3** *The announcement of the new 'compact' 2.4-litre sporting saloon, which was to prove of enormous importance to Jaguar sales (J C)*

2

3

remarkably quickly, and 1957 turned out to be a record year for output.

For the Tourist Trophy Meeting at Dundrod in 1954 Jaguar had produced a new 2½-litre version of the XK engine for two of the D-Types in an attempt to beat the handicappers, who were notoriously hard on the larger-engined cars. This engine reappeared in the completely new small saloon at the 1955 Earls Court Show.

Most important in 1955 was that this was the first production Jaguar to be built of unitary construction instead of traditional chassis and unstressed body construction of all the previous models. The D-Type had pioneered this type of construction for Jaguar, although in a rather different form, and it was certainly progress in the right direction.

The suspension was also different, being a coil spring and wishbone arrangement carried on a separate rubber-mounted subframe at the front and a rear suspension of trailing arms with cantilever semi-elliptic springs. New Lockheed Brakemaster-type hydraulic brakes were fitted all round, aided by servo assistance. Although this was something of an economy Jaguar, all the usual refinements such as walnu dash, controls and door cappings leather upholstery and so on wer included in the specification of the 2.4 which weighed some 27 cwt. Includin purchase tax, the 2.4 retailed at £1,29 for the special equipment model, whic included a heater, central armrest Jaguar mascot, vitreous enamelle manifolds and an electric clock. With th optional Laycock de Normanville over drive, the new model was just capable o the magic 100 mph, and would reac 60 mph from standstill in just 14 seconds. Viewed now, the car looks ver

1

lated, with its heavy pillars and full rear vheel spats, but at the time it was well eceived.

Towards the end of 1956, the Mark VIII appeared. Essentially similar o its predecessors, it had however a few mportant changes. It was fitted with a 3-Type-headed engine, which produced :10 bhp in this form with twin SU HD6 :arburettors. The B-Type without leference to the alphabet had been leveloped from the C-Type head. It 'aried in that it had an inlet valve angle of 45°. This resulted in a useful improvenent at the lower and middle rev ranges and slightly better overall performance. The extra power was somewhat offset by extra weight created by extra refinements – not that the three cigar lighters now fitted would have altered things appreciably, but it all added up.

Visually the Mark VIII differed from the Mark VII in four noticeable ways. First and foremost the dated V-type split windscreen was replaced by a curved one-piece item. The radiator grille was re-styled and a chrome strip followed the wing line down the sides, breaking up the slab effect. Finally the rear wheel spats were slightly cut away, again relieving those large slab sides. To further lighten the body style and follow fashion, the Mark VIIIs were generally finished in two-tone paint. More subtly,

1 *A one-piece windscreen, chrome surround to the radiator grille and side chrome body moulding immediately distinguish the Mark VIII (*National Motor Museum*)* **2** *More at home in its native country, where it was listed at £1,299, the 2.4 was very popular (*J C*)* **3** *A couple of 2.4 saloons involved in high-speed testing at the banked MIRA track, where they proved themselves capable of 100 mph in spite of their smaller engine (*N. Dewis*)*

2

3

a chromium-plated radiator surround and leaping Jaguar mascot adorned the front end.

This, then, was the Mark VIII climbing, as we would say today, up market, in an attempt to steal more and more customers from those two fine cars – Bentley and Rolls-Royce.

Late in 1956 the XKSS was created and became available at the beginning of the following year. It is said that due to lack of demand, amazingly enough, for production D-Types (substantiated in an advertisement in an *Autosport* of 1958 for a 'brand new D-Type nearly £1,000 under list price' – that is, £2,995!) several semi-completed cars were sitting around 'going rusty' when Sir William had the idea of clearing them by turning them into more civilized road cars. This is just one of several theories about the birth of the XKSS. Another was that it was to enable owners to compete in production car racing in the States; whereas Duncan Hamilton also claims the credit for the idea, having converted his first D, OKV 1, into a more refined car for its subsequent owner, the late 'Jumbo' Goddard.

The XKSS differed from an ordinary production D in various details. The centre division between driver and passenger was removed and a full width screen was provided. The headrest fairing was not fitted, but a door for the passenger was. The car sprouted front and rear quarter bumpers and a baggage rack mounted on the rear. Some extra trimming to the interior and rudimentary sidescreens and hood added a little to creature comforts.

Sadly, production was halted abruptly by the fire when only 16 XKSSs had been completed. Of these, 12 went to the States, two to Canada, one to Hong Kong and one remained in Britain. Two D-Types were converted by the factory to XKSSs, making a total of 18 all told.

In the aftermath of the near disastrous fire, the company launched a natural progression of the small car range – the 2.4 saloon but with the more usual 3442 cc version of the XK engine. This unit, with twin SU HD6 carburettors, high lift cams and a dual exhaust system, produced 210 bhp at 5,000 rpm and gave the 3.4, as it was known, a very useful performance. A genuine top speed of 120 mph, acceleration from standing start to 60 mph in 11.7 seconds and the standing quarter-mile in 17.9 seconds ensured that this was no sluggard and, as usual, kept Jaguar in front of its competitors.

Mechanical changes were made to take the extra weight and power of the larger engine. The engine mountings were modified, the suspension stiffened, and besides the larger grille and badges the main distinguishing feature between the two models was that the new addition had cut-away rear wheel spats. Manual, manual with overdrive, and Borg Warner automatic transmission were the options, and the last of these, although slower, still gave a brisk performance, with 60 mph reached in 11.2 seconds.

From September 1957 the 3.4's wider grille was adopted on the smaller version. Disc brakes could be ordered for both models from the beginning of 1958, and automatic transmission was now available on the smaller car as well.

An area in which both the XK120 and

1

2

XK140 had very definitely not been up :o standard was in braking. Fade had ılways been a problem and the few small nodifications over the years had made ittle real improvement. Some new levelopment was obviously needed, and ıere racing indisputedly improved the ɔreed. The XK150 appeared in May 1957, fitted with Dunlop disc brakes, which had first been used by Jaguar on the C-Type in the 1952 Mille Miglia. Disc brakes were a subject of considerable collaboration and development

1 *The rear spats of the Mark VIII had also come in for attention and contributed to the updating of the flagship* (J C) **2** *Members of Royalty were tempted to try the XKSS, the thinly disguised D-Type, here being sampled by Prince Michael of Kent with Norman Dewis* (N. Dewis) **3** *The XKSS was never a very photogenic car but looked much happier in the flesh* (J C) **4** *Prince Philip was not always content (as shown here) to be a passenger in an XKSS, and insisted on taking the controls when out of sight of his entourage* **5** *A curved windscreen, fixed side windows, two doors and a hood made the XKSS almost civilized* (J C)

5

BUBERY OWEN
WOLSELEY
STUDEBAKER
& PACKARD
DAIMLER
LINCOLN MERCURY
SIMCA VEDETTE
LADIES
TELEPHONES
CAR PARK
POST OFFICE
THE ROYAL SOCIETY FOR THE PREVENTION OF ACCIDENTS
34 LITRE

The fitting of the 3.4-litre engine to the smaller saloons gave them typical Jaguar performance (J C)

between Dunlop and Jaguar, and they played a great part in Jaguar's success on the track, particularly in long-distance racing, in which the Ferraris experienced brake fade with their then conventional drum brakes. These new Dunlop brakes were quoted as being 'the greatest single weapon which demolished the opposition' in the 1953 Vingt-Quatre Heures du Mans, when the three works C-Types finished in first, second and fourth places.

So, four years later, Jaguar introduced similar disc brakes on a production car in the shape of the XK150. Only fixed-head and drophead models were announced initially. The factory could not at this juncture cope with heavy re-tooling with other work on hand, so by cleverly compromising, they were able to use much of the 120/140 tooling. The exterior design of the car, although still very much an XK, was nevertheless significantly different from the XK120 and 140. The wing line, which had fallen and risen so dramatically on the earlier cars, in contrast was almost straight. The bonnet was wider with the addition of a fillet down the centre to reduce new tooling, making access to the engine a good deal easier. The windscreen was of one-piece design and the walnut dashboard was gone, replaced initially by aluminium and later by leather. The XK150 was a step further away from the out-and-out, wind-in-your-hair type of sports car introduced in 1948 and a step nearer to the ultra comfortable, ultra sophisticated grand touring car announced in 1961. This is not to decry the car, but is merely a reflection of the times and Jaguar's business policy.

The fixed-head illustrated here would have cost you £1,763 in the UK at the time of its announcement, just £123 more than the Special Equipment XK120 of 1953. So much for inflation. . . .

The interiors on both models benefited from the thinner doors, with shoulder room being increased by four inches, which made the car a lot more comfortable. Like the 140, the 150 had the two small occasional rear seats, but the heating layout, also similar to the

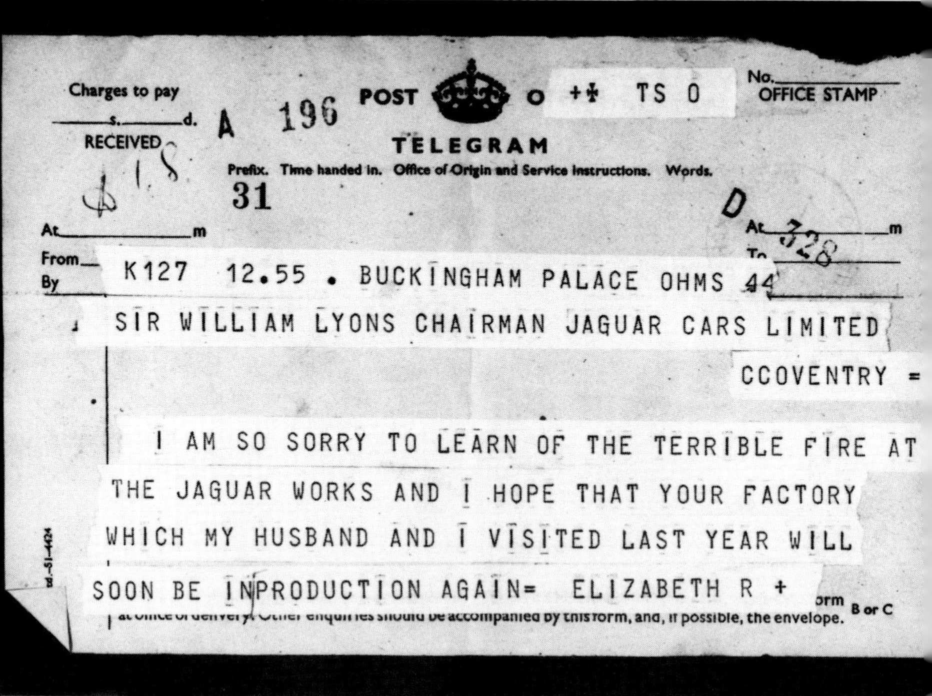

Charges to pay
s. d.
RECEIVED
POST OFFICE TELEGRAM
A 196 O +½ TS 0
No. OFFICE STAMP
Prefix. Time handed in. Office of Origin and Service Instructions. Words.
31
At m
From
By
K127 12.55 . BUCKINGHAM PALACE OHMS 44
SIR WILLIAM LYONS CHAIRMAN JAGUAR CARS LIMITED
CCOVENTRY =
I AM SO SORRY TO LEARN OF THE TERRIBLE FIRE AT THE JAGUAR WORKS AND I HOPE THAT YOUR FACTORY WHICH MY HUSBAND AND I VISITED LAST YEAR WILL SOON BE INPRODUCTION AGAIN= ELIZABETH R +

1

10 Downing Street
Whitehall

February 13, 1957.

Dear Sir William Lyons

I am indeed sorry to hear of the disaster which has struck your firm. Jaguar cars have been doing a magnificent job in our battle for exports and it is tragic luck that you have received this blow. But I am confident that with your customary energy and courage you will get production going in a very short time. I am the more encouraged in this belief by the accounts of the wonderful work done by your workers in helping to save the most important part of your factory.

Yours sincerely

Harold Macmillan

2

1, 2 *Such was the gravity of Jaguar's fire in postwar Britain that the disaster did not go unnoticed – even in the very highest circles* (J C) **3** *The sad remains of an XKSS and a damaged D-Type were among the casualties of the terrible fire of 1957* (J C) **4** *The Duke of Kent collects his new 3.4 from Lofty England, showing Royalty's preference for Jaguars* (National Motor Museum)

140, was now primitive in comparison with its competitors. In several respects, including the poor front seats (they were virtually a bench with their lack of support, which inhibited enthusiastic cornering), the XK150 was starting to show its age, in comparison with say, the 300 SL Mercedes Benz, the BMW 507 and the DB2/4 Aston Martin. But in fairness it must be stated that the basic XK design dated back to 1948 and its rivals were much more expensive. The basic XK150 DHC when introduced was, as stated, £1,783 (including purchase tax), whereas the 300 SL cost £4,651, the BMW £4,201 and the DB2/4 £2,889. Not surprisingly, the XK150 was the heaviest XK, at some 29½ cwt.

Initially two power options were available. The standard car fitted with the 140 engine gave 190 bhp. However, a special equipment model was also available and this was usually fitted with a 210 bhp engine. This latter unit was fitted with the B-Type head. This produced more torque rather than any great increase in power.

The XK150 was the first sports car to have the leaping Jaguar mascot to be officially offered as an optional extra.

At that time, American cars with sporting pretensions were being fitted with larger and larger V8 engines and, although they could not live with an XK when it came to cornering or stopping, some were faster in a straight line. Jaguar, conscious of this, and also of the performance of the various Italian exotica, made efforts to improve the 150.

To those visiting the Earls Court Motor Show in 1958, the new Mark IX, which succeeded the Mark VIII, seemed little different, but under the skin several important changes had been made to add yet more sophistication to Jaguar's flag-

1

2

3

ship. Not surprisingly, disc brakes were now fitted all round as standard and, better late than never, the large saloon was at last fitted with power assisted steering, a facility the Americans had been used to for many years on their own cars and a department in which they had criticized the Jaguars. The assistance was provided by a Hobourn-Eaton pump driven off the rear end of the generator, and a useful side effect was that this allowed the steering to be geared up to require only three-and-a-half turns from lock to lock, which suited those with sporting pretensions.

The Mark IX had an enlarged version of the faithful XK engine. Capacity was increased to 3.8 litres (3781 cc). It was the first time a production Jaguar had been fitted with this unit, although a number of examples had been on active service on the race tracks of the world in 1957 and 1958, particularly in the successful Lister sports racing cars.

This new engine improved the Mark IX's performance over that of the Mark VIII by a useful margin. Acceleration from 0 to 60 took just 11 seconds, and top speed was now more than 110 mph. Naturally, this had to be paid

1 *'Partners in Power' was the title of Dunlop's advertising campaign, featuring the disc brake and the XK150* (Dunlop) **2** *One of the legends of motoring journalism, John Bolster, about to try the new XK150 Fixed Head* (National Motor Museum) **3** *Now more of a grand touring car, the XK150, in Drophead guise, was a further step away from the out-and-out sports car introduced in 1948* (J C) **4** *The line of the XK150 DHC was pleasant enough but in practice the folded hood made rearward vision far from satisfactory* (J C) **5** *The S insignia below the front quarterlight identifies the triple carb, straight port-headed version of the XK150* (J C)

for, which meant £2,163 for the automatic version and a consumption of around 14 miles per gallon. This was not surprising when you consider that the car weighed 35 cwt.

A 1958 Jaguar sales brochure stated: 'For many the choice of a sports car means only one thing – a roadster; for this is the type of model from which all other sports cars have been derived and which alone possesses the classical sports car "line"'. And so, at the beginning of 1958, a Roadster XK150 was announced and with it came the optional S version.

The S had a further refined cylinder head, which had been developed by Harry Weslake. Known as the 'straight port head', because of the partial straightening of the ports, thereby increasing the flow of mixture at high revolutions, it was also fed by no less than three 2-inch SU carburettors. The engine was also fitted with a stronger clutch, lightened flywheel and lead-bronze bearings.

The result of this was that the car could reach 60 mph in 7.3 seconds and had a top speed of 133 mph. The Americans were quick to note that here was a car that was 'designed' to travel at these speeds, and consequently felt safer and more controllable.

In other respects the Roadster was similar to its predecessors, but with the 150 modifications. It was still a pure two-seater but less stark, which caused raised eyebrows amongst some diehards when they noted that it had wind-up windows.

Later, the S engine was offered on the fixed-head and drophead coupés as well. Searching for still more performance, Jaguar announced in 1960 that the 3.8-litre XK engine, which had already been offered in the Mark IX, was to be available in the XK150. The 3.8 could be had either with the B-Type head, or with the straight port head, whence it became known as the 3.8S. The latter was the ultimate production XK, with a power output of 265 bhp, giving a top speed of 136 mph. In acceleration, the car was actually slightly slower at lower speeds than the 3.4, but gained considerably nearer its maximum. The problem now, with its live rear axle, was to get the power down on the road efficiently.

Towards the end of 1959 Jaguar updated and improved the small saloon range and added a 3.8 model. Because these were called MK 2s, their predecessors have logically come to be known as MK 1s, although they were never officially so-called.

The appearance of the cars was considerably improved, as was vision, by a much larger wrap-around rear window and more slender pillars all round, accentuated by the use of chrome window frames. The facelift included a new grille with a central rib, faired-in sidelights on top of the wings, and horn grilles replaced by spotlights. The interior was also altered and equipped with a new and much superior instrument layout.

An increased rear track (by $3\frac{3}{4}$ inches) contributed to greater stability at speed, which had been a problem on earlie models; and the front suspension wa given a higher roll centre. Disc brake were no longer optional – they wer standard, and drivers found tha repeated use at high speeds caused n loss of stopping power or unevenness nor did it require an increase in peda pressure. A brake fluid level warnin light was added to the specification.

Three models of Mark 2s were avail able. These were the 2.4, with a

1

2

improved B-Type head producing 120 bhp at 5750 rpm; the 3.4, which continued to give 210 bhp; and the 3.8, first seen in the Mark IX a year earlier. This last-named engine had a 87 mm stroke compared with the 83 mm of the 3.4 unit. In producing 220 bhp it lost nothing of its smoothness or flexibility, and made useful gains in torque – 240 lb ft at 3000 rpm compared to the 3.4's 215 lb ft.

This largest-engined version, which also had a limited slip differential fitted, was an excellent high performance saloon with few if any rivals – certainly not at £1,842 15s 10d for the manual with overdrive. Its vivid acceleration in third gear on wet roads could easily cause wheelspin. And in competition 3.8s clocked up a prodigious number of wins in subsequent years.

A contemporary motoring magazine came to the conclusion that the 3.4 offered an 'outstanding combination of

1, 2 *More for export than for home consumption, the XK150 Roadster, as it has come to be known, was officially called the Open. Unlike the Drophead, it had an uninterrupted rear view* (National Motor Museum) **3** *Like the XK150, the Mark IX was the final version of a theme started at the beginning of the decade* (J C) **4** *The disc brakes fitted to the Mark IX were now even more necessary with the 3.8 engine* (J C)

speed, refinement and true driving ease.' The Mark 2 models were to be of immense business importance to the company because while the competition models provided the image and publicity, the sports models the excitement, and the executive range the prestige, the compact saloons contributed most to the turnover, stability and steadily increasing profits of the company. Nearly 100,000 Mark 2s were produced. It should be remembered that while many firms have had periods of difficulty and a great many famous names have disappeared for good, Jaguar have sailed happily on with full order books at all times, or at least until the late 1970s.

The order placed by the Nottingham police for a number of Brough Superiors fitted with Swallow sidecars set a precedent that was to continue to the present day. Postwar Mark V saloons were favoured by several forces, but it was the later Mark 1 and 2 saloons that proved particularly popular, especially for patrolling the newly opened M1, which at that time had no speed limit at all. These saloons provided the police with just what they required: reliability and performance at a sensible price.

The XK engines were put to good use in vehicles other than cars. They were used in Dennis fire tenders and in various military vehicles, and required little modification. When fitted to the Alvis Scorpion and Scimitar armoured vehicles a military-type distributor was fitted, the heads were altered in the exhaust valve region to meet the requirements of heavily leaded fuel, and a Solex carburettor replaced the normal SUs.

It is no coincidence that many of the original Jaguar distributors and agents prospered and continue to prosper

1 *The Kent family remained loyal to the marque, Prince Michael being among the very first to aquire a new Mark 2 saloon (*J C*)* **2** *The chrome side window frames of the Mark 2 did much to modernize and lighten the smaller saloon range (*J C*)*

1

2

JAGUAR IN COLOUR

Colour is, and always has been, important to Jaguar, being a vital ingredient of style and individuality – both important features in the Jaguar story. The Swallows and SSs were boldly coloured, distinguishing them in an otherwise dour era. The post-war colour range was extensive and incorporated a selection of highly effective 'coalescent' finishes. Today, new colours are helping to popularize the current range.

*R.M.V. 'Soapy' Sutton about to achieve 132 mph in a standard XK120 Roadster with the normal screens removed (*JC*)*

1

2

3

1, 2 *The central slat to the radiator distinguishes the Mark II Austin Seven Swallow saloon, fitted here with ships' ventilators atop the scuttle and clearly showing the 'peak'* **3** *The SS I Tourer offered SS style with open motoring and four seats (*J C*)* **4** *The delightful rakish lines of the SS 100, together with its lively performance combine to make the model one of the most revered of all sports cars (*J C*)* **5** *From left to right: a $1\frac{1}{2}$ sv Saloon, an SS 100 and an SS 1 Saloon (*J C*)*

4

1

2

3

4

6

1 A high proportion of SS100s survive today allowing their owners to still appreciate their style and respectable turn of speed **2** *Jaguar's first new postwar model, the Mark V, clearly shows the prewar influence but was to be the final manifestation of that line (*J C*)* **3** *The rear threequarter vision must have been appalling on the Mark V Drophead Coupé, but there is no denying the car had style (*J C*)* **4** *Leslie Johnson's XK120 before the start of the 1949 Production Car Race at Silverstone, the car's racing début and first of countless victories* **5** *Another view of Johnson's car which was in fact the second left-hand drive XK120 built and later converted to right-hand drive (*J C*)* **6** *That XKs were popular for club racing can clearly be seen from this Le Mans-type start (*J C*)*

5

1 *The first C-Type, XKC 001, photographed at the factory before being shipped over to Dundrod where Johnson and Rolt finished third behind team mates Moss and Walker in the 1951 TT* (J C) **2** *Here Moss pilots C-Type XKC 053, which he shared at Le Mans in 1953 with Peter Walker, to second place behind the similar car of Rolt and Hamilton* (J C) **3** *Norman Dewis is congratulated by Lofty England after recording the amazing speed of 172 mph at Jabbeke in Belgium* (J C) **4, 5** *This D-Type, XKC 404, was one of the original trio of works Ds for 1954, the model's first victory being the 12-hour event at Rheims. OKV 3 continues to be actively and successfully campaigned today in international historic racing by Martin Morris* (J C)

1

2

OKV 3

OKV 3

1

1 *This Mark VII Saloon belonged to H.M. The Queen Mother for a number of years and was partially updated to Mark VIII specification, which explains the one-piece windscreen* (J C) **2** *An XK150 Roadster with an interesting selection of optional extras displayed behind* (J C) **3** *The XK150 'S' Roadster in 3.8-litre right-hand-drive form was produced in very limited quantities* (J C) **4** *Jaguar's Mark 2 Saloon was an obvious choice for the police to use for patrolling Britain's first motorway, the M1* (J C) **5** *The Mark 2 Saloons were immensely popular and remained in production from 1959 to 1967* (J C)

2

POLICE

3251 HP

1 *A very early pre-production E-Type on test at MIRA. Note the absence of chrome rims to the headlamp glasses (*J C*)* **2** *Even today, over 20 years on, the Series One E-Type remains a sensationally exciting shape (*J C*)* **3** *This E-Type underbonnet view shows the designers' success in carrying out Lyons' original dictum – that the engine was to be a glamorous one (*J C*)* **4** *A Series One Fixed Head E-Type, displaying its pure, clean lines (*J C*)* **5** *Sir William Lyons' dollar-earning flagship for the 1960s, the Mark X, in front of his magnificent home, Wappenbury Hall (*J C*)*

1

2

4

1 *The independent rear suspension unit developed for the E-Type and Mark X, and subsequently used on the interim S-Type and 420, and more latterly the XJ range (*J C*)* **2** *An early example of an XJ6, the range that was to carry Jaguar's fortunes through the 1970s and early 1980s (*J C*)* **3** *The Series Three E-Type introduced the brilliant V12 engine to the world (*J C*)* **4** *The closed V12E was offered only as a '2+2' and like its open sister was introduced in 1971 (*J C*)* **5** *The XJ, seen here in Series Two form, continued the long partnership between the police forces of Great Britain and the Jaguar company (*J C*)*

1

2

TRW 920J

UHU 700N

1 *The sensational XJ13: exciting because of its styling, an extension of the C/D, D, E theme, its mid-engine configuration, and four-overhead cam V12, but sad because it was never seen in action (*J C*)* **2, 3, 4, 5** *Interesting styling exercises, about which little is known, but perhaps a replacement for the E-Type was being contemplated (*J C*)* **6** *The XJS undergoing wind-tunnel tests (*J C*)* **7** *Often described as a 'plumber's nightmare', the underbonnet area of the XJS with the V12 engine installed left little room to spare (*J C*)* **8** *The XJS, whilst not a sports car, was a 'grand touring' car in the sense of the phrase's original definition – a car capable of carrying its occupants at high speed and in great comfort (*J C*)*

1

2

3

4

5

7

1

2

3

1 The Group 44 XJ27 with the V12 engine installed amidships was surely designed and built with Le Mans ultimately in mind and was entered for the '84 race (J C) 2 The two most important men in Jaguar history. On the right, Sir William Lyons, founder, former Chairman and Life President, with Chief Executive, John Egan (J C) 3 A typical Jaguar motor show stand of the early 1950s showing a Mark VII Saloon surrounded by an XK120 Roadster, Fixed Head and two Dropheads (J C)

ılthough a few have been lost to the narque, thanks to British Leyland's ranchise policy.

From the start, Brown and Mallalieu, or whom Lyons had worked briefly, and Parkers, were agents for Blackpool and Manchester respectively. When Lyons visited Birmingham, he met up with the partners of P. J. Evans, later known for a while as Broad Street Motors. They were appointed distributors for the Midlands. Henlys became distributors for he south, and over the years have leveloped into a very large group.

The later distributors and dealers list ounds like a Who's Who of famous garages. It included H. R. Owen, Mann Eggerton, Blue Star Garages, Lex Garages, Rowland Smith, Rossleigh, Martin Walter, Charles Follett, Sydney Marcus, Harold Radford and Kennings, to name but a few. Others, not surprisingly, were involved in Jaguar competition history. These included Sammy Newsome, John Coombs, HW Motors, Ian Appleyard, Dickie Attwood, Archie Scott Brown, Reg Mansbridge, Robin Sturgess, the Stewarts (Jimmy and Jackie), Dick Protheroe and many others who, like the above, were either the owners or the sons of owners. Many, such as Rothwell and Milbourne of Malvern and Ritchies of Glasgow, have been loyal to the make for many years.

1 *It is hard to believe that the Mark 2 could carry all this equipment and still be ideal for high-speed pursuits* (J C) **2** *The mobile theme taken up by Jaguar's largest distributor, although in a rather more sophisticated form* (National Motor Museum)

7 THE E ARRIVES

Sensation returned to the Jaguar image with the unleashing of the dramatic E-Type sports car and later the prestigious Mark X. Export took on an even greater significance for the company and its products demanded and received renewed respect. The smaller saloons continued with refinements and added variations and brought practical Jaguar motoring to a wider audience.

1

2

3

There was no denying that in 1961 the XK150, good car though it was, was becoming a little dated in comparison with some of the competition, and certainly by Jaguar standards. Sir William and his team had been working on a replacement since 1957. They had learned a lot from the racing D-Types and in March, a new Jaguar, the sensational E-Type, was unleashed on the world. The world was stunned, standards changed overnight, competitors reeled, and Jaguars were back at the forefront of sports car design.

Later that year the factory introduced a completely new, and at the time very dramatic, large saloon car, the Mark X. Both these models received considerable acclaim across the Atlantic, earning very many dollars for Britain, and both reflected the increasing technical merit of Jaguar products.

The E-Type, arguably the most famous sports car of all time, has certainly become a household name and a synonym for speed. Introduced at the Geneva Motor Show in March 1961, there were two versions available – the open two-seater Roadster and the two-seater Fixed Head Coupé.

Closely derived from the D-Type of Le Mans fame, and again a perfect example of racing improving the breed, the E

1 *Sir William Lyons poses with his ultimate sports car at Geneva in March 1961 (*Autocar*)* **2** *E1A, the first step down the road to the legendary E-Type (*N. Dewis*)* **3, 5, 6** *Rather nearer to its final form, this car seen on test from a variety of angles is presumed to be the one affectionately known in the factory as 'the pop rivet special' (*N. Dewis*)* **4** *The famous road-test Fixed Head, 9600 HP, that set the 150 mph legend when tested by most of the major motoring journals (*Autocar*)*

6

shared a similar monocoque centre section to which was attached sub-frames front and rear to carry the front suspension, steering and engine, and the rear independent suspension and differential, respectively. Fitted with the XK engine in 3.8 form, triple carburettors and four-speed box, allied to the overall lightness of the car, it boasted a staggering performance for its day.

The biggest deviation from the D-Type theme was in the rear suspension. Independent suspension had been tried on a D, and was further developed on E2A, the second prototype, which was raced for a while by Briggs Cunningham (including Le Mans in 1960), but this was the first time it had been used on a production Jaguar. Certainly it contributed greatly to the car's high standard of ride, roadholding and overall performance. The car's braking, only just about adequate for the performance, featured disc brakes all round, inboard at the rear. The only department in which the 3.8s could be, and have been, seriously criticized was the gearbox. It was heavy and very slow, and slightly marred the overall pleasure of the performance.

The E-Type, unlike the XK150, was not available with either overdrive or automatic transmission – there was simply no room.

The original Jaguar brochure opened with the sentence, 'No more famous background can be found anywhere than that which lies behind the Jaguar E-Type GT (Grand Touring) Models'. It then went on to name the models as 'the Jaguar E-Type GT Fixed-Head Coupé and the Jaguar E-Type GT Open Two-Seater'. Interestingly, like the original XK120, correctly entitled the Open Two-Seater Super Sports, but which

1

2

3

4

has come to be known simply as the Roadster, somewhere along the line the GT was also dropped from the E-Type's title. The same brochure concluded its preamble with the modest claim that 'the Jaguar E-Type GT is, in truth, the most advanced sports car in the world'. Few, if any, would have disagreed with this statement.

For 1962 a few minor revisions were made. Notably the bonnet, which had been opened by a T handle inserted into catches on the exterior, was now released by internal catches. The bonnet louvres, which on early cars had been a separate panel let into the centre section, became a one-piece integral pressing. And a little more leg room was achieved by means of a 'dish' in the floor. A slight indentation in the rear bulkhead allowed the seats to be moved slightly further back for taller occupants.

Prices stood at £1,830 for the Roadster and £1,954 for the fixed-head. Production figures for the 3.8s were 7,827 and 7,669 for the Roadster and Fixed-head, respectively.

The E-Type's performance is well known. It had a top speed of around 150 mph, and a 0–60 mph time of 6.9 seconds, with consumption in the area of 16–20 mpg, depending on whether you were pressing or pottering.

What is not quite so well known is that the original fixed-head road-test car (registered 9600 HP and now owned by

1 *The press were riveted by the sports car's dramatic appeal (*Autocar*)* **2** *In this seat you could travel at 150 mph – still a legal speed in 1961 (*Autocar*)* **3** *Lyons's original dictum was fulfilled: the engine was indeed a glamorous piece of machinery (*Autocar*)* **4** *Sir William Lyons enters 77 RW, the original open-road test car at a Jaguar Silver Jubilee gathering (*National Motor Museum*)*

the author), had one or two novel features. 'The car that created the 150 mph legend' as it has been described, had an aluminium rear tailgate in place of the normal steel, perspex windows, and an interesting engine.

The E-Type in standard form would not, in fact, do a genuine 150 mph, and later models have always appeared slower by comparison with those original figures. But this is mere quibbling today when we have blanket speed limits and dense traffic, to say nothing of most of the drivers. The E-Type was and is a very fast car.

Following closely on the announcement of the E-Type came the new 'top of the range' big saloon, the Mark X. Replacing the Mark IX, the new car was a considerable step forward rather than a mere updating of a concept. Gone was the chassis, the construction being of a monocoque type in line with other models. The styling, while bearing some family resemblance, was generally rather different. Designed with the US market in mind, it was extremely large, and wider than any other British car of the period. The reverse slant to the fron end, incorporating the fashionable twir headlamps, was distinctive.

The Mark X shared a good deal witl the sports car. Most notable was the full independent rear suspension, whic gave the car a high standard of ride an cornering. This incorporated on eacl side a lower transverse tubular linl pivoted at the wheel carrier and the sul frame adjacent to the differential casin and, above this, a half shaft universall jointed at each end. These served t locate the wheel in a transverse plane

1

2

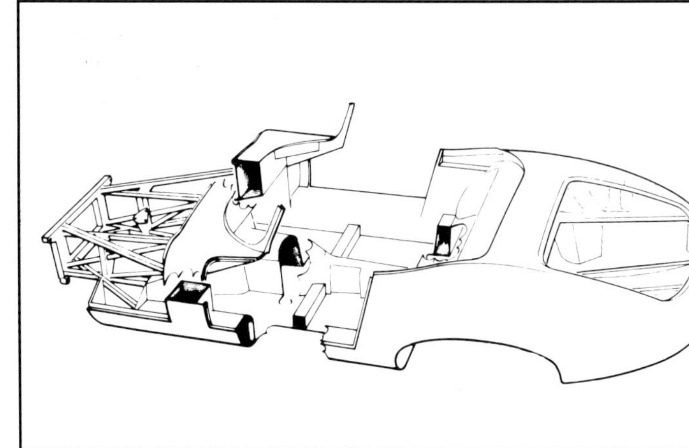

3

Longitudinal location was provided by rubber mountings, which located the sub-assembly to the body structure and by radius arms between the lower links and the mounting points on the body. Twin-coil springs, each enclosing a telescopic hydraulic damper, provided the suspension medium.

The front suspension was also an independent arrangement and incorporated double wishbones and coil springs with telescopic shock absorbers. An anti-roll bar was located between the lower wishbones. The complete assembly, together with steering gear, was mounted on a separate sub-frame, which was itself located in the body by rubber mountings.

To power this very big car (the boot alone had a capacity of 27 cu ft) which weighed 35 cwt, Jaguar provided the 3.8 engine with three SU carburettors, first

1 *Few Es found their way on to the home market in 1961, but this car, now registered XVE 1, the first of the line, went to Henlys as their demonstration model* **2** *For the first time, an open Two-Seater Jaguar could be had with a hard top – the external bonnet release lock of the very early cars is visible (*J C*)* **3** *The monocoque and sub-frame construction of the E-Type (*J C*)* **4** *There was a strong family resemblance between the E-Type and its racing forbear, the D* **5** *The dramatic new flagship, the Mark X, enabled Lyons to 'steal' another Motor Show* **6** *The counter-balanced forward-hinging bonnet of the Mark X was an innovation for a Jaguar saloon, as were the 14-in diameter wheels (*National Motor Museum*)*

6

The original E-Type had a clean, aerodynamic shape (J C)

seen in the XK150S and used subsequently in the E-Type. With an output of 265 bhp at 5,500 rpm, it gave the Mark X a lively, if not staggering performance and an average fuel consumption of 16–18 mpg, although fast motoring could reduce this to 12 mpg. Fuel for this thirsty animal was carried in two 10-gallon tanks in the rear wings.

Most cars were fitted with automatic transmission, but manual transmission could still be specified, with or without overdrive. In practice, few manual versions were produced. Comfort was of a high order, with leather seats, and very full instrumentation laid out in the usual Jaguar tradition, surrounded by walnut. Electrically operated windows became an optional extra for the first time on a Jaguar.

Needless to say, the car was a great success. It caused a sensation at the Earls Court Motor Show, and earned yet more dollars for Britain. As usual, journalists vied with each other to think of more and greater superlatives. Denis Holmes, writing in the *Daily Mail*, after claiming 'It's new; it's astounding; it's unequalled', reached his verdict with 'The chauffeur will have to beg to be allowed to drive'.

With regard to the dollars earned, Jaguar were able to announce, 'At the end of an all-day meeting between Jaguar and their American and Canadian distributors, orders were placed to a total of 63 million dollars (£22½ million). This represents a doubling of the demand for Jaguar in the North American continent, and results from the widening of the scope of the Jaguar range following the introduction of the new Mark X saloon. This order will stretch the production capacity of the factory to its utmost even allowing for the expansion programme already in hand. A strict delivery schedule has been written into the contract and orders not executed on time will be automatically cancelled. Sir William said at Earls Court recently "This order represents the biggest challenge in our career, and will need every effort by both the management and men in order to meet it. It presents an opportunity which we may never have again to establish an unassailable position in the American market. I am bound to say that the delivery dates which the distributors have specified leave no margin for any delays".'

Unfortunately, a few teething problems marred the early life of the Mark X to a small degree. There were problems with the radiators and there occurred a deterioration of the body to sub-frame rubbers.

Announced in 1963 and produced from 1964 until 1968, the S-Type saloon was a compromise heralded by the factory as 'the latest development of one of the world's most successful cars'. To call it a compromise may sound a little unfair because it was a very good car, inheriting a number of the better features from the models that sired it. It was basically a Mark II in shape and specification, with the following exceptions. Most importantly, it had the E-Type's independent rear suspension, which gave considerably improved roadholding and comfort over the Mark IIs, and for an increased outlay of only £200. It had the longer rear-end boot treatment along the lines of the Mark X, giving an increase of 7 cu ft over the Mark II, and was slightly re-styled at the front. A flatter roof line and changes to the rear seats gave more head and leg room in the back.

Available in both 3.4 and 3.8 forms (but not with the 2.4 unit because of the S-Type's extra three cwt) and priced at £1,669, and £1,758 for the larger-

1

2

engined version, the S-Type was a happy compromise that sold steadily. It moved one motoring magazine to comment that the car was 'outstanding value for money, combining effortless high speed cruising for four people with really reassuring road manners'. The ride, as a result of the independent rear suspension was, they said, 'probably unexcelled by any other European car'.

In 1963 Massimo, an Italian firm, announced a 12-plug head for the E-Type. Little more is known about these heads or whether many were produced, but the concern claimed 'a very considerable increase in power output', aided no doubt by the three Webers they also fitted!

In 1964 the E-Type matured a stage further with two major alterations and several smaller ones. Firstly, the 3.8 engine was replaced by a 4.2 version of the trusty old XK engine. The 4235 cc capacity was achieved by re-designing the block and resulted not in any greater power output, still quoted at 265 bhp

1 *An imposing car from any angle, the Mark X's bulbous sides now date the car somewhat (J C)* **2** *The S-Type, available in both 3.4 and 3.8 form, was a marriage of Mark 2 and Mark X ideas (J C)* **3** *Apart from adding the 4.2 engine in 1964, Jaguar silenced the only major criticism of the earlier E-Type with an all-synchromesh gearbox (J C)* **4** *Still looking as fresh as it did at its introduction, the Series I 4.2 had not yet suffered the constraints that were to come (J C)* **5** *The 4.2 Mark X was notable for its new Marles Varamatic Bendix power-steering, unique to Jaguar (J C)*

5

1

2

(gross) at 5,500 rpm (as opposed to 5,400 for the 3.8), but did achieve a significant improvement in torque, rising from 260 lbs/ft at 4,000 rpm to 283 lbs/ft at the same engine speed.

The other major alteration was long overdue, namely the fitting of an all-synchromesh gearbox. The slightly increased power which this absorbed was more than compensated for by the increased satisfaction on the road.

Outwardly the cars remained virtually unchanged, the only obvious way of identifying the new models being the small '4.2' on the rear. Inside, the cars had much improved seats with the bright aluminium fascia and transmission tunnel giving way to leather-finished items introduced in late 1963, and an alternator was also fitted. Gone was the old Kelsey-Hayes bellows-type brake servo, to be replaced by a Lockheed vacuum booster.

These changes resulted in a slight increase in weight, and prices increased to £1,896 for the open car and £2,032 for its closed sister. Production figures were 9,548 and 7,770 respectively.

In 1964 the Mark X was further developed with the enlarged engine of 4.2 litres being fitted, as with the E-Type. The exterior remained unchanged apart from the addition of a small '4.2' on the boot lid.

A new improved Borg-Warner automatic transmission was offered and the power steering came in for some attention with the adoption of the Marles Varamatic variable rate system. A manual version was still offered and the new gearbox, again shared with the sports car, was a considerable improvement. This was appreciated by motoring writers, who had been less than generous in their comments on the previous box. They also approved of the more efficient and controllable heating system.

With these all-round improvements the Mark X continued to be in great demand both at home and across the Atlantic. The detail improvements enabled it to be described as 'a large power-

1 *The larger-engined Mark X was enhanced by revised automatic transmission and Dunlop radial tyres (J C)* **2** *The 2+2 version of the E-Type lost a little of its line. This early car does not carry the chrome trim below the side window that was normally a distinguishing feature of the 2+2 (J C)* **3** *The 420 saloon's engine was fitted with two carburettors instead of three, and as a result developed 20 bhp less than its larger sister (J C)* **4** *The central slat to the radiator and chrome side moulding clearly distinguish this 420G (J C)*

ful, roadworthy and luxurious car able to transport five people and their luggage in great comfort over long distances'.

Interestingly, the Jaguar brochure for this model stated that 'the latest techniques and equipment for rustproofing and painting are utilized to ensure maximum body life and higher standards of finish.' This anti-corrosion treatment was hardly effective (as can be judged today), indicating that the body engineers still had a fair amount to learn in this department. The ravages of time have been far from kind to the Jaguar ranges of the 1950s and 1960s, as anyone restoring an example today is only too well aware.

Typical of Jaguar development activities, Norman Dewis, Jaguar chief tester, surprised his fellow motorists with an indecently fast Mark X. The bonnet was well and truly locked down, but it was rumoured that underneath, the engine had twice the normal number of cylinders. Could this be a sign of things to come?

The year 1966 saw a new model which further extended the E-Type's role and therefore its market. Until then, the E had been strictly a two-seater with the various limitations that had imposed on owners with families or other commitments. The introduction of a 2 + 2 model enabled two extra adults to be conveyed for short distances or two children to be carried indefinitely. This model, being rather longer, made the option of automatic transmission possible and increased demand from the United States, where a high proportion of motorists favour, if not expect such an option.

The 2 + 2 was lengthened by 9 inches, the roofline raised by 2 inches, the doors extended by $8\frac{1}{2}$ inches, the frontal area increased by 5% and the weight by 2 cwt. Mechanically, the car was only altered to care for increased weight by minor changes to spring rates and damper rates, etc. Inevitably, performance suffered, with a top speed for the automatic version of only 136.2 mph, and it took 1.6 seconds longer to reach 60 mph from standstill.

In July 1966, much to many people's surprise, Sir William Lyons and Sir George Harriman of BMC (British Motor Corporation) announced that their companies would merge to form British Motor Holdings (BMH). Autonomy would be maintained and strength and security gained from unity. That was the theory.

Although it may seem a harsh judgement the 420, announced at Earls Court in October 1966, was merely an interim model. It was a further compromise along S-Type lines and can simply be described as an S-Type with Mark X frontal treatment and the 4.2 engine. It served to show the way the Lyons theme was developing with looks somewhat akin to the completely new saloon range that was to follow a couple of years later. But in spite of this it was not a bad car. One leading motoring magazine considered it to be 'a saloon that for a combination of speed, comfort and safety is as good as any in the world, regardless of cost'.

The following optional extras could be specified when ordering a 420: foglamps at £15 9s 9d the pair, power-assisted steering at £67 12s 1d, air-conditioning for £280 5s 0d, an electrically heated rear window for £18 8s 9d, a laminated windscreen at £6 9s 0d and wire spoke wheels at £43 0s 5d silver painted, or £95 5s 2d chromium plated. All prices included purchase tax. The list concluded: 'Every effort is made to meet individual requirements. The above list details extras most often requested'.

The large saloon displayed at Earls Court in 1966 was no longer a Mark X but a 420G. You would have been forgiven for being a little perplexed because outwardly they looked identical apart from a chrome strip down either side of the 420G with a small flasher unit at the forward end of it, a bolder central slat to the grille and the substitution of small grilles for the previously fitted spotlights. The price was increased by £81 to £2,237.

Inside the car, as a sop to the safety

1

2

protagonists who were beginning to be heard, the dash top sprouted a thick padded roll, which incorporated a clock. The seats were slightly altered in an effort to improve lateral support, because the front seat passengers had tended to slide around on the vast straight seats (which resembled those of the XKs) during spirited cornering. Otherwise the car was unchanged.

A limousine model had also been produced during 1966. It had first been seen in Mark X form and continued in 420G guise. This varied only in that it had a division between the front and rear seats, a cocktail cabinet and larger picnic tables. It was claimed that the rear seat was one of the widest available and 'provides ample lounging room for three full size adults'.

Jaguar were able to announce in 1966 that exports of postwar models to America now topped 200 million dollars in total value.

The Mark IIs became 240s and 340s in 1967, and the 3.8 model was deleted from the range. The so-called new models differed little from their predecessors. The concept was becoming a little dated by then, but it should be remembered that the original 2.4 was announced in 1955, so it is hardly surprising.

In the interests of economy, hide seats were discontinued, and the Mark II's spotlights gave way to small circular grilles first seen on the Mark Is. In an attempt to modernize the model slightly, the previous rather heavy double curvature bumpers were replaced by redesigned single slimline examples.

Meanwhile the evolution of the E-Type took a further step with the introduction in October 1968 of the Series Two version of the Roadster, the fixed-head and the 2+2. The majority of changes were to the exteriors and were largely dictated by US Federal Safety Regulations, the bane of exciting cars.

The front was altered, with a larger bonnet intake (for the optional air-conditioning unit), raised wrap-round

1 *The Mark X/420G limousine carried added refinements for extra privacy (*J C*)* **2** *Made for only one year, the 340 was now the only other version because the 3.8-litre engine had been dropped (*J C*)* **3** *The new slimmer bumpers of the 240 and 340 were an attempt to prolong the life of the ageing Mark 2s*

1

2

bumpers, the disposal of the headlamp covers (actually carried out a little earlier on what are known as Series $1\frac{1}{2}$ cars), the movement of the headlamps up and forward so that they protruded above the bonnet, breaking the smooth line, and larger combined sidelamp and flasher units. The rear suffered in a similar way, with again larger and higher wrap-round bumpers, much larger rear lamp units and a square number plate. The car also carried additional flashers on front and rear wings and de-eared spinners. In line with the regulations, the interior had minor changes, including the obligatory rocker switches.

Braking was considerably improved with a Girling system with three pistons at the front and two at the rear. Power steering was now an optional extra. To satisfy the US emission controls, models exported to those shores were fitted with Stromberg carburettors. This, together with the extra weight and the less clean shape, resulted in the performance being by E-Type standards pedestrian, with a top speed somewhere in the region of 125 mph.

The 2+2 shared all the changes of its sisters plus one further alteration. The rake of the large windscreen was increased by 7 degrees by moving the base forward virtually to the bonnet line. This helped the looks of the car a little, where it had previously looked too severe.

As with the XK150s in their day, the competition was catching up with Jaguar again, particularly in the States. Cars such as the De Tomaso Pantera, a combination of European chassis and styling with the brute force of an American Ford V8 engine, were showing up the E-type in stifled form. As a result, Jaguar needed to take a large technical and image-boosting step forward once more to keep ahead. And they had the answer up their sleeve.

1, 2, 3 *The Series II E-Type had to satisfy new safety regulations – and suffered in appearance as a result. Further sops to the safety protagonists were, amongst other things, the recessed door handles and rocker switches. The rear end suffered similarly, with its over-heavy treatment now dictated by regulations* (J C)

8 THE XJ RANGE

Unquestionable technical merit yielded high praise for the saloon range that replaced the ageing models in 1968, and the technical qualities were further enhanced by the sophisticated V12 engine fitted first to the final progression of the faithful E-Type. Its successor had initially to endure uniquely dark days for the company, but improved quality control, technical refinement, decisive inspired leadership and at last a replacement engine for the brilliant yet ageing XK unit, pulled the company through and spelt a highly optimistic future.

1

In 1968 Jaguar announced the XJ range, and again the company had a world-beater. It was and continues to be, more than 15 years later, widely acclaimed. The faithful and constantly developed XK engine was still used, and this was supplemented by the new V12 engine first used in the ultimate progression of the E-Type series in 1971. This sensational engine was very highly regarded and gave Jaguar the technical image boost just as the XK engine had done in 1948. A combination of pleasant styling and first-class engineering made these cars clearly the best Jaguars yet. The story has turned full circle, because just as the SSIs in their day were unkindly nicknamed 'the Bentleys of Wardour Street', so in the 1970s the motoring press compared the XJ12 with the Rolls Royce Silver Shadow ('The best in the world'), and asked, Which is the better? This was a tribute indeed to Sir William, who could see in the XJ the fulfilment of his original ambitions, from manufacturing stylish sidecars to arguably the finest production car in the world.

That the XJ range, still in production today in a revised form, was and continues to be a world-beater is too recent to need emphasising. As usual, the motoring press competed to express its extreme appreciation of the virtues of the new model introduced in 1968. One magazine stated that it was concerned

1, 2 *Jaguar had another winner in the XJ6 and the major problem, as it had so often been, was meeting demand. Unquestionable technical merit impressed the motoring world, apart from the usual Lyons gift for styling seen once again in the XJ6* **3** *The XK engine was 20 years old but nobody would have known because it powered the XJ6 in silky silence* **4** *The V12 E-Type's major exterior change from the earlier models was the addition of a grille to enclose the famous mouth (*J C*)*

4

that, being frustrated like many customers at having to wait months before being able to acquire an XJ6, the critics might make that frustration even greater for the customer by lengthening the waiting list still more. They went on to say that they felt that the car was 'closer to overall perfection than any other luxury car we have yet tested, regardless of the price'. It is interesting to note how often that phrase 'regardless of price' has been applied to Jaguar's products.

The new model, known simply as the XJ6, was offered with a choice of engines, either the familiar 4.2-litre or a new 2.8-litre version of the trusty XK design. Interestingly, Jaguar hinted of things to come when they made the significant statement in their press release at the launch, that 'Within the next two years, it is intended to introduce new additional power units into the range. The XJ6 has been designed to accept them and, therefore, in due course we plan to introduce further XJ models featuring the new engines'.

Construction, as with the previous models, was of the pressed steel monocoque variety with a massive central platform and stout scuttle structure. By thus concentrating the strength in the lower half of the body, the need for thick screen pillars and rear quarters was avoided.

Innovations as far as Jaguars were concerned were the provision of anti-dive geometry in the newly designed front suspension, and the fitting of new low profile, radial-ply tyres developed in conjunction with Dunlop. Another first for a Jaguar saloon was the choice of rack-and-pinion steering, power-assisted on the 4.2 and 2.8 de luxe models, and with that assistance as an optional extra on the basic 2.8-litre. Rear suspension was the usual independent set-up in a separate sub-frame.

Many safety features were built into the original design, including a collapsible steering column and independent front and rear hydraulic circuits for the Girling disc brakes, fitted all round.

Priced at £2,314 for the 4.2 manual gearbox with overdrive, and £1,797 for the basic 2.8-litre, these cars in the true Lyons tradition offered unbeatable value for money. The $32\frac{1}{2}$ cwt did not help consumption, which averaged 15–17 mpg, but this was before the petrol rises and the economic depression.

In 1971 a manual 4.2 XJ6 supplemented the automatic versions. The car was somewhat quicker, but not many were made because automatic transmission was more popular for a car of this type. The clutch on this model was particularly heavy, which did not endear it to drivers.

Braking for most purposes was more than satisfactory but the author recalls a high speed run in the late 1970s with Mike Barker of the Midland Motor Museum in their C-Type and Gordon Benbow in a Series 1 XJ6. I was in my XK140 Roadster. After about 30 miles of pretty hard motoring Benbow had to slacken off because the XJ's brakes were fading. This was ironic when you think of the XK's poor old drum brakes, but it was probably explained by the fact that it was rather easier to scrub off speed by lateral cornering in the XK, and by the XJ's greater weight.

In mid-1969, the man whose contribution had been second in importance only to that of 'the Old Man', left the stage. Bill Heynes had achieved the rank of Vice-Chairman (Engineering) when he retired, and his country had rightly rewarded him by honouring him with a CBE for his services to Jaguar.

For reasons of prestige, apart from its technical advantage, it was felt by Jaguar

1

2

n the 1960s that they must have a glamorous and technically sophisticated engine to power the models of the 1970s and beyond, in the same way as they felt n the 1940s when they designed the six-cylinder.

The engineering brief for the new engine called for a design offering a greater degree of smoothness, silence and flexibility than the six, coupled to power and torque outputs which, in the V12 engine's basic form, were required to be at least as good as the highest figures achieved by the XK unit. These in their most advanced competition form produced approximately 325 bhp, and the V12 production engine came close to achieving this.

Like the development of the XK before it, the best ever achieved by the previous engine was made the starting point for the new. The parallels are fascinating.

The cylinders of the 12 were set in a V formation at an angle of 60 degrees with, in production, a single overhead cam per bank, although early versions designed with racing in mind had twin

1 *The greater length helped to offset the more cluttered front and rear treatment of the Series III E-Types (J C)* **2** *The centre-mounted V badge is the clue to the engine fitted to the XJ6 derivation, the XJ12 (J C)* **3** *Like the XK150 in relation to the XK120, the V12E was thought by many to be a completely new model and is better regarded as such (J C)* **4** *To answer one of the few criticisms of the model, later XJs were available in long wheelbase form, adding four inches of legroom and $1\frac{1}{2}$ cwt (J C)* **5** *Continuing their long association, several police forces acquired squads of XJs, here in revised Series II form, for patrolling motorways (J C)*

overhead cams per bank. But eventually, cost, under bonnet space, noise and complexity decided the issue.

A bore of 90 mm and stroke of 70 mm gave a capacity of 5343 cc (326 cu in) which, with a compression ratio of 9:1, gave a gross power output of 314 bhp at 6200 rpm. An innovation at that time was the fitting of Lucas transistorized ignition. Aluminium was used extensively – the block was of aluminium in order to save weight.

Jaguar stated: 'In this engine we have provided quite a substantial reserve for capacity increases should the need arise (and) . . . it is relatively easy to derive a substantially smaller engine from the design'.

An entirely separate engine assembly line was set up for the production of the V12 engine and this was situated at the Radford plant. Total tooling-up costs were in the region of £3 million, with equipment for machining the heads accounting for some £700,000, and this at 1970 prices.

Assembly was carried out on an electrically driven 52-stage track, designed for ease of working, with vigorous inspection standards and final bench-testing before transfer to Browns Lane. Justifying the heavy capital outlay, Jaguar optimistically stated that 'the installation is geared to produce a future optimum of 1,000 major power unit components per 80-hour (two week) shift, and it is possible to utilize the equipment to produce alternative capacities and specifications. . . .'

It was no secret that Jaguar were working on a V12 engine in the 1960s. Its origins can be traced back to 1955 and the racing days when it was considered that more power would be needed to keep ahead of the opposition, so some drawings were made. As with the XK engine, it was designed for the saloon range and was fitted in the sports car as a 'production test bed'. It was destined for the XJ saloon, but first it was to be offered in the E-Type.

Externally the Series Three E differed considerably from its earlier sisters. A grille appeared at the front to replace the open mouth. Wheel arches were slightly flared to accommodate wider tyres and increased track. Only a 2+2 and a version known simply as 'open' were offered and the latter, although it had only two seats, shared the longer wheelbase chassis of the 2+2. This change allowed the open model also to be available with automatic transmission. The previously optional power steering was now standard and very necessary because of the extra weight of these cars – 28.8 cwt for the Roadster and 29.5 cwt for the 2+2, an increase of nearly a quarter over the original Series One model. However, in spite of this weight disadvantage and the further cluttered shape, the V12 engine produced performance figures comparable to the early cars and were a considerable improvement on the interim models.

Prices of the new models were in the true Jaguar tradition, the Roadster being listed at £3,123 and the closed version at £3,369. It is interesting to note in these inflation-beset days that the Roadster had increased in price from 1961 to 1971, a period of 10 years, by £1,293 or some 70%.

Inevitably, the XJ12 eventually arrived, and the poor motoring journalists, who only four years earlier had used all the superlatives they could beg, borrow or steal now had to redouble their efforts. Some simply asked 'The most luxurious car in the world?', while others stated it as a fact.

The V12 engine, already tried and

1

2

:ested in the Series Three E-Type was, ıs mentioned, always destined for the ;aloon. Even so, the installation was not ın easy one. It was a very large engine, ·specially with all the ancillaries dictated ɔy comfort and regulations, and this left ɔrecious little space to spare in the :ngine compartment. This resulted in :he problem of how to dissipate surplus ıeat. A series of modifications to the :ooling system, including a newly lesigned cross-flow radiator, solved the ɔroblem. Such was the thoroughness of the Jaguar engineers that a fan was fitted to cool the battery in case of high ambient temperatures.

Apart from the fitting of ventilated front discs, a different grille and different badges, the car was largely the same as the six-cylinder version.

The V12 engine gave the XJ12 a top speed of around 140 mph and meant that this was the fastest production four-

*1 The interior of the Series II XJs demonstrated that in spite of modern ergonomics Jaguar still favoured good traditional quality (*J C*) 2 The two-door pillarless treatment of the XJCs particularly suited the model (*P. Porter*) 3 With both side windows down, the Coupé was a most attractive, stylish and almost sporting car (*P. Porter*) 4 The XJS undergoing the extensive development testing that is an ever increasing part of a new model's gestation period (*N. Dewis*) 5 The newly announced XJS leads a parade of racing cars round the streets of Birmingham – no prize for guessing which of Jaguar's distributors it belonged to (*J C*)*

4

seater in the world in 1972. And the price for this privilege was just £700 more than that of the XJ6, making a total of £3,672.

In the pre-petrol crisis world, demand was very heavy and the consumption, which could be as high as 12 mpg, was irrelevant to many people. The waiting list was so long that Jaguar's own advertising used cartoons with captions such as 'Welcome to the first XJ12 up our street'.

Jaguar had acquired Daimler in 1960 and now indulged in that questionable art of badge engineering. In late 1972 a Vanden Plas version of the Daimler XJ12 known as the Double Six, was announced, at a cost of £5,363. The significance of this model was that the wheelbase was increased by 4 inches. The two Jaguar saloons became available in this long-wheelbase form, whence they were known as XJ6Ls and XJ12Ls.

In 1972 Sir William Lyons, at the age of 71, bowed out, some 50 years after those humble beginnings and ardent ambitions. Lofty England stepped up to become his successor as Chairman and Chief Executive, and Lyons took the honorary title of President.

Towards the end of 1973, at the Frankfurt Show, the factory announced new improved versions of the popular XJs, and these were christened the Series Two.

The V12-engined saloon was no longer available on the short 108.8-inch wheelbase, but only in the 112.8-inch body. The 2.8 was dropped from the home market and from then on was available only in certain continental markets where taxation or insurance made it viable. The 4.2 version now had the ventilated front discs and sculptured wheels fitted previously to the V12 only. The new cars also featured a single-tube oil cooler, a brake pressure differential warning actuator and a revised exhaust system that eliminated the flexible pipe section. Anti-intrusion barriers welded into the doors provided protection designed to meet legal requirements anywhere in the world. Improved sound deadening on the front bulkhead included the elimination of all grommets, a full-width asbestos sheet attached to the front of the bulkhead and the interior face of the bulkhead; and the footwells and transmission hump had a moulded sound-absorbing material that consisted of Olfield bitumen and cotton felt and Hardura PVC foam with a loaded PVC surface. Dedication indeed to the continuous quest for more silence!

The interior came in for considerable re-vamping, with a complete new dash layout. The confusing tumble or rocker switches had given way to a more modern push-push variety and there was a general re-grouping of instruments.

Externally the new models had detail changes that made them look quite different. In order to satisfy North American safety requirements for 1974 the front bumpers were raised 4 inches to a height of 16 inches. This necessitated a new smaller grille above the bumper and below a rectangular grille with horizontal bars that concealed the new oil cooler. The rear came in for less attention. Lighting for the rear number plate was moved from below the plate to above it and formed a single chromium-plated metal casting with the boot lid lock. A new heating system, the fitting of power steering, a heated rear window and

laminated windscreen as standard completed the main changes. The long wheelbase versions also had electrically operated windows and head restraints as standard equipment.

The various improvements enabled Jaguar's to keep abreast if not ahead of their main rivals, notably the BMWs and Mercedes of Germany, and executives everywhere continued to covet the model. But at the factory, management and workforce were going through turbulent times. A year after Lofty took the helm, Lord Stokes appointed outsider Geoffrey Robinson as Managing Director and several months later, to the disappointment of many, Lofty retired. In mid-1975 Robinson resigned over the infamous Ryder Report and its suicidal proposed submersion of Jaguar's autonomy and very identity.

The range for the early 1980s posed alongside an RAF Vulcan (J C)

At the same time as Jaguar released the Series Two saloons, they also announced a new model, the XJ Coupé. Mechanically, and in most other respects identical to the saloons, the Coupés had one major difference and it was a striking one – namely, the cars had two doors instead of four, and pillarless construction. The doors were enlarged by 8 inches and the side windows, electrically operated, could be completely lowered from sight thus giving a very sporting appearance. This was further enhanced by a neat vinyl roof.

Based on the short wheelbase body, the car was still a full four-seater, with access gained to the rear by tilting the front seats forward.

To make good any torsional stiffness lost through the omission of the central pillars, the rear quarters were increased in thickness and reinforcing beams inside the body were provided. In fact, production did not begin for a couple of years because of teething problems with torsional rigidity, and effective operation and sealing of the windows.

The company's advertising claimed, 'That comparatively rare person, in whom a spirit of individuality remains strong, will find freedom of expression in the dignified, exclusive yet vivacious performance of these superb cars'. This was quite reminiscent of the early Swallow days. Certainly the Coupés were more distinctive than the saloons, and their relative exclusivity was ensured when the company decided to concentrate on the long wheelbase saloons. This decision killed off the Coupé, the V12 version of which, with fuel injection, was capable of around 145 mph.

In the mid-1970s the company introduced fuel injection for the V12 XJs for reasons of economy. For similar reasons the old faithful 3.4 litre, although in rather different form, was re-introduced for the four-door saloons. In 1975 the 3.4, at £4,795, cost some £342 less than the 4.2-litre.

When the XJS was at last announced in 1975, the critics were divided in their opinions. This was a unique situation for a Jaguar sports car. But there was a problem – it was no longer a sports car. Some complained that it lacked the dramatic appeal and sporting pretensions of the XK120 and E-Type before it; others claimed that this did not matter, that it was a high-speed luxury car appropriate to the 1970s and 1980s. Nobody tried to call it a sports car, and nobody denied that it was technically superb. Really it was the culmination of a trend away from the wind-in-the-hair extrovert sports car first seen in 1948, a trend continued by the less sporting XK150 to the more sophisticated E-Type to the even more sophisticated V12E. What would those diehards who complained at the wind-up windows of the XK150 Roadster say when they observed that those of the XJS were electrically powered?

Technically the XJS was closely based on the XJ saloons, with similar monocoque construction, and the same front anti-dive suspension and independent rear suspension, although both front and rear were stiffened with a new rear anti-roll bar of 14 mm and a larger front one. This resulted in a firmer ride more in keeping with the car's sporting pretensions instead of the slightly spongy feel of the saloons. The power steering had been geared down for more feel, taking just three turns from lock to lock. The engine was the familiar V12 unit but fitted with Lucas fuel injection. For transmission you had a choice between the usual manual four-speed box or Borg Warner three-speed Model 12 automatic. Braking was very adequately taken care of by the 11.18-inch diameter ventilated discs with four piston calipers at the front and the inboard 10.38-inch diameter discs with two piston calipers at the rear, along with vacuum assistance and dual hydraulic lines.

People were also divided on their views of the styling. Certainly the influence of the safety fanatics could be detected immediately and clearly in the heavy black impact-absorbing rubber bumpers, which were mounted to Menasco struts that acted like shock absorbers to withstand the compulsory 5 mph impact. Creature comforts were well cared for, with air conditioning standard, and reclining front seats with separate squab centre section for lateral support. The interior was trimmed throughout in flame-retardant materials and no fewer than 18 mechanical and safety functions were monitored through a bank of coloured lights. Red lights indicated major faults, such as brake failure, and amber lights denoted secondary faults such as brake bulb failure.

With a top speed of 150 mph, the ability to cruise in almost total silence, with a consumption of 15–18 mph, excellent road manners and great safety, the XJS could well be, technically at least, what the advertising script claimed – 'The pinnacle of Jaguar evolution'.

Jaguar testing and development has long been extensive and exhaustive, but never more so than today, with the increased complexity of the many worldwide technical regulations to be met. This has resulted in a greatly extended embryonic period, to the frustration of many. Overseeing the physical testing of all cars since 1952 has been Norman Dewis. He has been involved with both racing development and the production side, to say nothing of the high-speed publicity runs and the driving of the racing models to the events.

Early in 1979 the XJ saloons were given a facelift – in other words, the body was modernized. This was achieved by subtle restyling rather than major surgery. The roofline was raised at the rear to give a lighter look and more headroom, and the inevitable heavy black bumpers incorporating front indicators and rear fog lights were grafted on. The design, re-tooling and production costs amounted to £7 million, which is interesting to compare with the £1 million needed some 24 years earlier to completely develop the small 2.4-litre saloon. Uniquely for the company, an outside design studio was retained by Jaguar to advise on this XJ revamping.

New flush door handles, an optional wash/wipe device for the headlamps, the outer of which now incorporated the sidelights, a new grille, a new rear light cluster and number plate lamp were all features of the Series III.

On the mechnical side, the 5.3 remained unchanged, but the 3.4 and 4.2 could be had with a five-speed gear box (ex-Rover suitably modified) or the Model 65 Borg Warner automatic box. The 4.2 was fitted with Lucas fuel injection and the inlet valves were increased to $\frac{7}{8}$-in, the same as the D-Type head, resulting in a useful increase of 30 bhp.

The interior had deep-pile carpeting

1

2

or greater luxury and even greater sound deadening, and one magazine commented, 'Other cars are quiet but Jaguars (and Rolls Royces) are silent'. Front seats came in for attention with the adoption of an adjustable lumbar support system, and an optional extra was the electrical height control of the driver's seat. Other optional extras included a cruise control device, an electrically operated sliding roof, and many more minor items.

At the time of writing, the basic XJ 3.4 saloon costs just £11,188, or £11,017–10s-od more than the original basic Swallow Seven two-seater; but one can hardly compare the two, to put it mildly. While the Swallow was a good little car of its type, the latest XJs can openly compete for all-round excellence with any other car produced today. It is a matter of from simply being imitators of style in the company's youth to becoming leaders in mass production engineering sophistication in the fullness of maturity; and from offering a naïve but pretentious product to providing a 'full-bodied' one, par excellence.

In 1980 a saviour in the form of John Egan arrived as full-time Chairman of Jaguar. During the period following Robinson's departure in 1975, the company had been 'barely' run by a series of committees with several virtually nominal figureheads. Apart from Bob Knight's brief spell, all the chiefs had been outside men.

Egan, however, brought fresh hope. Here was a man with respect for the past and belief in the future. He brought an intelligent coherence once more to Jaguar's existence. Gradually he tackled the problems, concentrating particularly on the wayward quality that had sadly been a part of the 1970s. He rekindled Jaguar's autonomy, rebuilding the image. And slowly the company returned to profitability. Exports, particularly to the States, began to match the results of earlier years, and the company again became Jaguar Cars Ltd.

A few days after his 80th birthday in 1981, Sir William was quoted in the *Birmingham Evening Mail* as believing that Jaguar's only hope was to 'go it alone'. Still President, he stated, 'I think it's a very good idea to sever Jaguar links with BL. Jaguar should be a separate entity, and I think Sir Michael Edwardes is leaning that way.' Lyons believed that

1 *Perhaps the best view of the high-speed executive express that visibly reflected American safety regulations in its front and rear treatment* (J C) **2** *The controversial rear end treatment of the XJS. Whatever the styling, you could travel at high speed and in unprecedented silence* (P. Porter) **3** *Soon after its introduction with automatic transmission only, the XJS was offered with the V12E's four-speed manual box* (J C)

top management at Jaguar was in good hands. 'They are making a success of things and restoring the company to its original standing.'

In November 1982 *The Sunday Times Business News* published a full-page article detailing Jaguar's recovery and dramatically increased export business with the headline: 'Jaguar Climbs out of the Pit'.

Fuel economy became increasingly important in the late 1970s and early 1980s, following the petrol crises and dramatic increase in prices. Mindful of this, Jaguar engineers turned their attention to reducing consumption of the six- and particularly the thirsty twelve-cylinder engines. As a result, they introduced a new version of the latter engine termed the HE, the initials standing for High Efficiency.

The modifications consisted of the application of the Fireball combustion chamber principles, invented by Swiss engineer Michael May, and used for the first time in a production car. Jaguar engineers had spent five years perfecting the May principles to suit the demands of a production engine. The secret of the design lay in the split-level combustion chamber arrangement, with the inlet valve set into a shallow collecting zone and the exhaust valve set higher up within the bath-tub type combustion chamber, which also housed the sparking plug. A swirl-inducing ramped channel connected the two areas. The resulting swirl characteristics ensured rapid and complete burning of very lean fuel mixtures.

The XJS HE, now claimed by the factory to be the fastest automatic transmission production car in the world, reminds us of the claims made for the XK120 when it was announced. The mechanical changes improved consumption from an urban figure of 12.7 mpg to 15.6 mpg, with 27 mpg at 56 mph and 22.5 mpg at 75 mph, with a top speed of 155 mph. The coupé had a different bumper arrangement, with an element of chrome lightening the effect of the large black appendages. A tapered double coach-line distinguished the side aspect of the new model and the bonnet sprouted a flat circular jaguar's head badge not dissimilar from that of the XK120s. Five-spoke alloy wheels, veneer fascia, extensive use of Connolly hide, Dunlop D7 tyres on $6\frac{1}{2}$-in rims and a new Philips 990 micro-computer stereo radio-cassette unit completed the updating.

The twelve-cylinder saloons received the same HE engine, the title therefore of XJ12 HE, alloy wheels, electrically operated sun roof and door mirrors, and a headlamp wash/wipe.

On 12th October 1983 Jaguar announced additional new models but, in the longer term of more significance, a new engine. The model had been rumoured for some time and was an open version of the XJS, the Cabriolet. The engine was the new six-cylinder, the eventual successor to the perennial XK unit some 35 years after the latter's first appearance. The XK continued to power the 3.4 and 4.2 saloons but the new engine was obviously intended for the XJ40 range of saloons scheduled for introduction in the mid-1980s. Interestingly, following established Jaguar practice, the new engine was introduced in a low-volume sports model first.

Plans to produce a three-quarter scale 3.5 litre V-8 version of the V12 engine, thereby enabling the unit to be built on the same production line thus keeping capital investment to a minimum, had to be dropped when it was found the engine in this form suffered from secondary out of balance forces. The fitting of a balance shaft cured this but precluded using the V12 lines, so it was abandoned in 1972.

Further experiments included the building of half a V12 in the form of a slant six cylinder. This engine of 2.6 litre capacity was found to be lacking in power and any increase in stroke made the engine too high to machine on the V12 line. Next a V12 unit was built, spurred by competitive thoughts, in 1972 with four-valves-per-cylinder heads based mainly on the great experience Wally Hassan and Harry Mundy had gained whilst building Grand Prix engines at Coventry Climax with this configuration. This engine which developed 630 bhp was not persevered with as competition ideas receded but the lessons learnt were to prove valuable.

In 1974 a four-valve version of the XK unit was built with a view to producing the unit on the existing six-cylinder engine facilities. The resultant 3.8 unit proved satisfactory but the constraints of a 30-year-old design, the need for greater economy and the work being

1

2

done with the May head on the V12 all counted against this plan. Thus the major decision had to be taken to lay down a new engine assembly line. This decision gave the designers a much freer hand, the only proviso being that the May head for the new engine had to be able to be produced on the same line as the V12 May head; thus the cylinder bore centres necessarily had to be the same on both engines. It was decided to build two versions of the six, one with a May head for economy and one with the 24 valve head for performance.

The new engine introduced in the XJS had a bore and stroke of 91 × 92 mm, giving a capacity of 3,590 cc, and with the 24 valve head developed 225 bhp at 5,300 rpm. This compared with the XK unit in 1980's guise giving 162 bhp in 3.4 litre form and 205 bhp in 4.2 litre form.

The massively ribbed, deep skirted aluminium block extended well below the crankshaft which ran in seven main bearings. The extensive use of light alloys meant that the new AJ6 (as it was called) was some 25–30 per cent lighter than the 4.2 XK unit. Other features included Lucas digital electronic fuel injection and a Lucas constant energy ignition system. For reasons of height the engine was inclined 15° from the vertical.

The AJ6 was developed by Hassan and Mundy's successors, Jim Randle, Engineering Director, and Trevor Crisp, Chief Engineer (Power Units). The capital investment required amounted to over £21 million with £6 million alone for the new transfer line for machining the cylinder blocks. Jaguar stated that this new line had a capacity of one thousand blocks a week.

As stated the AJ6 engine was introduced in the XJS in both the Coupé version and the new open Cabriolet. The 3.6 Coupé was priced at £19,250 and offered with 5 speed Getrag box giving the car a much more sporty feel. Top speed was 145 mph and 60 mph could be reached in 7.6 seconds. Consumption was said to be around 25 mpg for mixed driving conditions. Externally the main distinguishing feature was a reprofiled bonnet with long bulge to accommodate the new engine.

The Cabriolet shared the same mechanical specification and new bonnet. Internally, however, there were changes for this was a two seater with the area behind the seats used for luggage storage only. The XJ-SC 3.6 was not an open car in the traditional 'convertible' sense but had what has come to be known as a 'targa' top. Fixed cantrails and a centre bar allowed various options. The roof consisted of two interlocking panels which could be removed together or singly, being stored in an envelope in the boot. To the rear could be fitted either a hard-top type heated rear window or a double-skinned folding hood.

The new shape – though more pleasing to the eye – was obviously a little less efficient in the windstream as top speed was quoted at the slightly more modest figure of 142 mph.

If prewar models were criticized for 'more show than go', then you might be tempted to say that in the 1980s the roles have been reversed. You might further surmise that the vintage Jaguar years were the 1950s and early 1960s, which were the years of the exciting XKs and early E-Types, and the years of the race- and rally-winning saloons. It is no coincidence that those were the competition years. But with the present management committed to excellence, marque individuality and pride, the situation is clearly improving. The return to the race tracks, the improvements in styling and quality control, and the strengthening of the financial status all bode well for the future of Jaguar Cars. After the dark days of the later 1970s, we can feel confident that the company now has the team to build on the great history of a magnificent past.

The all-important product is right again, the image has returned and there are exciting developments afoot.

1 *Subtle restyling and the inevitable impact bumpers easily distinguish the Series III XJ models, belying the fact that the XJ was more than 10 years old* (J C) **2** *The later XJS HE was improved in appearance by the addition of a thin chromed top bumper blade and 'cheese-cutter' wheels* (J C) **3** *Marque name changes led to partial dropping of the Daimler name and this model, the V12 Saloon, became the Jaguar Sovereign HE* (J C) **4** *The XJSC 3.6 Cabriolet with both roof panels removed and the optional folding rear window down is a more stylish machine. Note the discreet bonnet bulge to accommodate the new AJ6 engine* (J C)

9 JAGUAR COMPETES

Whilst it is clearly impossible to quantify Jaguar's competitive successes in business terms, what is undeniable is that the undoubted technical attributes of their cars came to the public awareness more swiftly and comprehensively than any advertising campaign could have achieved. International success against such august names as Ferrari and Mercedes-Benz demanded international press coverage, and greatly enhanced Britain's name abroad.

1

The importance of Jaguar's involvement in competition and its successes cannot be over-stressed. The many international awards and the resulting glory have done more for the company's image, and therefore sales, than possibly any other single factor. Furthermore, technical development has been aided and greatly hastened by racing and rallying. Many features first used on competition models were later seen on production cars. Above all, the five Le Mans wins in the 1950s ensured that the name of Jaguar would be a legendary one in racing history.

After the war, many of the competition successes were masterminded by another legendary Jaguar figure – Lofty England, who modestly described himself after the first Le Mans victory in 1951 as 'the sort of unofficial Competition Manager'.

As Lofty said recently, 'We realized that competition was the best way of producing publicity.' He went on to state that 'winning Le Mans didn't cost us much. I'd be surprised if it cost more than £15,000, but it was most important for it helped our sales in America and this helped our steel quota, which was a problem in the early postwar years.'

In 1976, a journalist writing about the

1 *Outside assistance needed for this SS I Tourer seen tackling a hill during the Barnstaple Trial* (National Motor Museum) **2** *An Austin 7 Swallow Two-Seater receiving assistance on a test hill in Yorkshire, which formed part of the London–Edinburgh trial* (J C) **3** *Old No 8, the famous SS 100, rounds the top S at Shelsley in September 1937, driven by Sammy Newsome in a best time of 45.52 secs.* **4** *M. J. Holloway is assisted to the famous start line at Shelsley. Old No 8 had by this time shed its wings and lights.* **5** *Sammy Newsome has a 'moment' between the esses at the same meeting in May 1938, later recording 43.98 secs* (W. Gibbs, Midland Automobile Club)

3

5

D-Type in the American magazine *Road and Track*, did so under the heading of 'The World's Longest-Running Advert'.

In the early days of SS competition activity, the cars won more prizes for styling and concours d'elegance rather than for performance. They initially did little to refute their image of 'more show than go', but gradually notable exceptions caught the motoring public's attention. The open SS Is distinguished themselves on several occasions, and Lyons entered a team of three SS I Tourers in the International Alpine Trial of 1933. Mechnical troubles asserted themselves, but the following year a team award was gained.

The 2½-litre SS100 was the first model to achieve any real distinction in competition circles, and thereby helped to allay the comments about more looks than performance. One of the first successes was the winning of the Glacier Cup and, in effect, overall victory by Tom and Elsie Wisdom in the 1936 International Alpine Trials – an event that was to figure several times in the future.

The car used was owned by the factory and, because of its chassis number – 18008 – came to be known as 'Old No 8'. It was driven in hill-climbs, particularly Shelsley Walsh, by Sammy Newsome (the Coventry garage and theatre owner), and in rallies and races such as Brooklands, by Tom Wisdom.

In 1937 a works team of 2½-litre SS100s was entered in the RAC Rally. The cars were driven by Tom Wisdom, Ernest Rankin (head of publicity), E. H. Jacob, and the Hon Brian Lewis. They won the Manufacturer's Team Prize but first place overall went to the privately entered SS100 of Jack Harrop, with Bob Taylor navigating.

In the Welsh Rally in July, Jacob put up the best overall performance, and the

1

2

3

1 *In spite of foul weather, Harrop and Taylor in their privately entered SS 100 beat not only the works team but also all other competitors to finish first in the RAC Rally of 1937* (R. Taylor) **2** *Jack Harrop (left), later killed in the war, and Bob Taylor his navigator, prepare for the 1939 RAC Rally, having notched up a second win in 1938* (R. Taylor) **3** *Harrop and Taylor hurling the SS 100 through the driving test stage of the 1939 RAC Rally* (R. Taylor) **4** *In spite of victories in 1937 and 1938, the best Harrop and Taylor could manage in 1939 was eighth* (R. Taylor) **5** *An SS 100 leaves the Great West Road starting point with 500 miles ahead through the Welsh countryside*

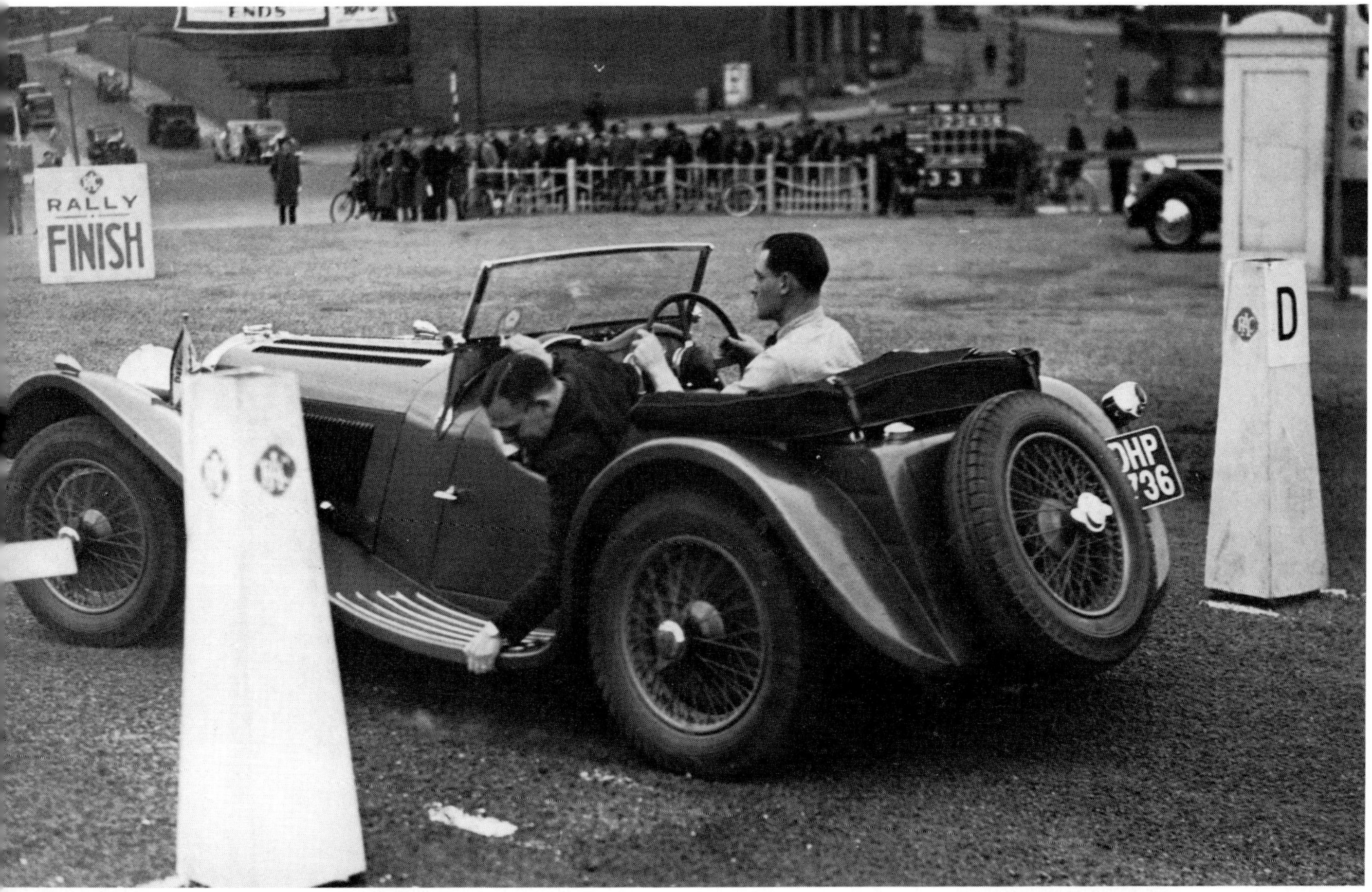
RALLY
FINISH
D
DHP 736

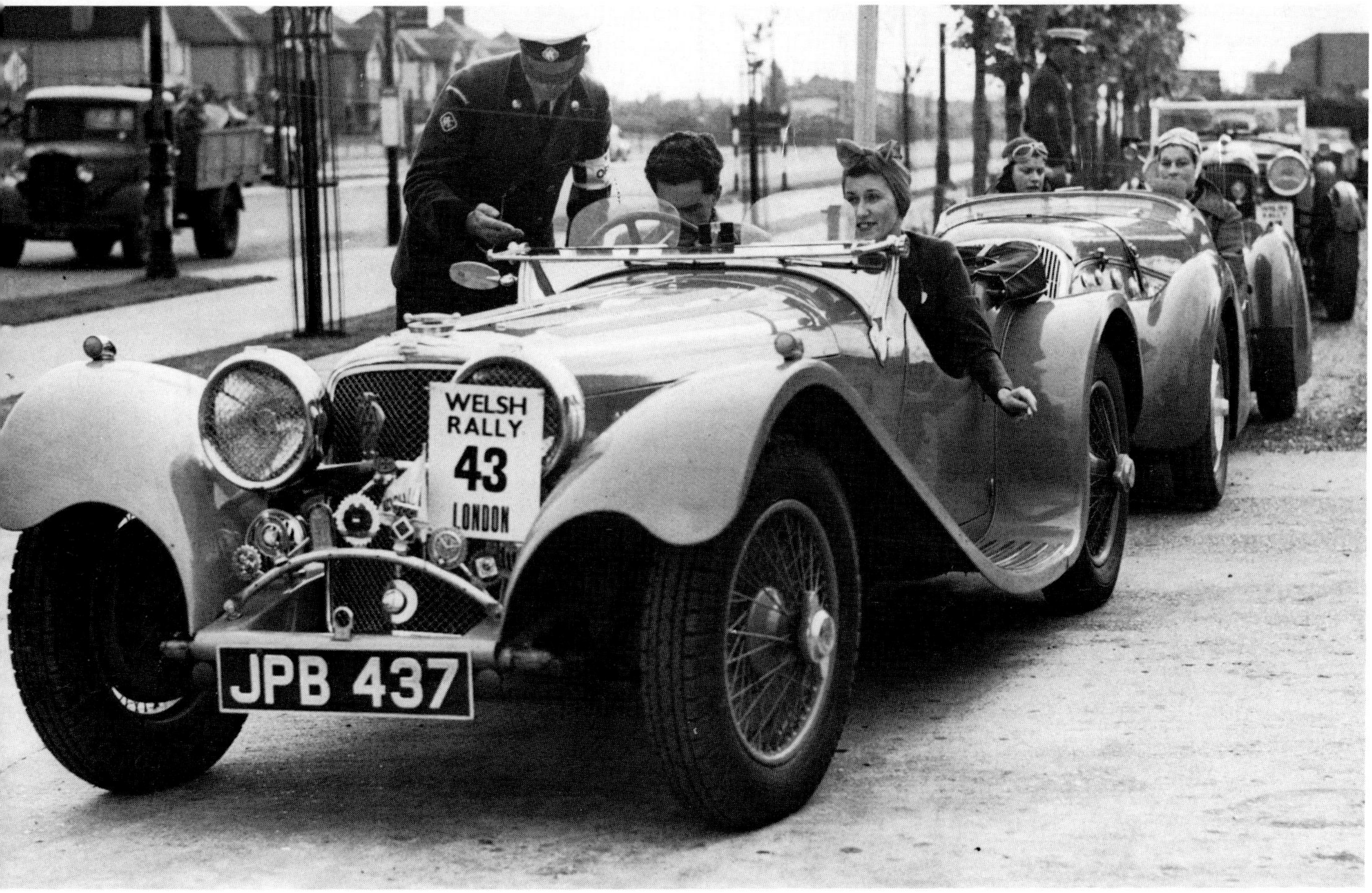
WELSH
RALLY
43
LONDON
JPB 437

team again took the Manufacturer's Prize. In October they won a 'long handicap' at Brooklands with Old No 8 now fitted with an experimental 3½-litre engine.

After their fine drive in the 1937 RAC Rally, Lyons invited Harrop and Taylor to join the works team for the 1938 event. It proved to be a good move because they brought the 3½-litre 100 to a win in the class for open cars over 15 hp. The same class was won by Mrs V. E. M. Hetherington in the Welsh Rally in her SS100, and Tom Wisdom added a first in class on the Paris–Nice Trial to his list of successes.

At an SS Car Club meeting at Donington, Lyons himself won a race for 'trade drivers', from Newsome and Bill Heynes, all in SS100s, of course.

Tom Wisdom writing some years later, stated 'As catalogued, the 3½-litre engine developed 125 bhp on a compression ratio of 7 to 1. What was its ultimate limit, modified for racing? Hassan and designer Heynes decided to find out. Encouraged by the robust bottom end they experimented with magneto ignition and a compression like a seizure, 17 to 1 to be exact. This called for methano fuels.'

Old No 8, fitted with this 3½-litr engine, continued to gain successes a Shelsley and elsewhere, and the engin was further developed until it was givin at least 160 bhp which, interestingly, wa the target set for the XK engine.

Apart from the engine, the body an chassis also came in for attention. Th body gradually shed parts such as wing and lights. It was given a neat, rounde tail, and the chassis was extensively dril

1

2

3

ed for lightness.

The 1939 rally season opened with the famous Monte Carlo event and Jack Harrop, this time without Bob Taylor, who had business commitments, finished in tenth equal place, driving an SS Jaguar 3½-litre saloon. The RAC Rally resulted in a second in class for Newsome (not in Old No 8) and second in the club Team Prize. In June, Newsome gave Old No 8 its final prewar run at Shelsley, and a win in the 3001–5000 cc unsupercharged class was his reward. Finally, in July W. C. N. Norton won his class, and Miss V. Watson took the Ladies' Prize (open cars) in the Welsh Rally.

SS100s continued to be used to great effect after the war and some of the model's greatest successes were gained in the early postwar years. With a dearth of race meetings, 100s distinguished themselves at sprints and hill-climbs, such as Prescott, Naish House and Shelsley. Ian Appleyard finished third in the 1947 Alpine Trial in his nine-year-old car. A year later he was back with a 'new' car – LNW 100, this being a car

1 *Bertie Bradnack waiting to commence a run at Shelsey Walsh (*B. Bradnack*)* **2** *W. H. Waring and W. H. Wadham take their 3½-litre Mark V through the braking and acceleration tests to finish ninth in the 1951 Monte – they were placed second in the Coachwork 'Comfort' Contest afterwards (*Fox*)* **3** *Leslie Johnson wrestles manfully and effectively with the XK 120 in its first race at Silverstone in 1949* **4** *Following Bira's spin Johnson took over the lead and Walker (seen here) finished second in the Production Car Race* **5** *Johnson crosses the finishing line to give the XK 120 its maiden victory first time out (*J C*)*

stored throughout the war and first registered in 1947. With this famous car he won a coveted Coupe des Alpes, and finished first in class. In 1949 Appleyard finished second in the Tulip Rally with the same car.

Various other examples continued to be rallied in the early 1950s with some success, but gradually other makes and, in particular, the XK120s, took over.

In spite of the Jabbeke demonstration of the XK120, a show of the XK's racing prowess was needed. An opportunity presented itself when the BRDC announced that they would be holding a production car race at their August 1949 meeting. Before committing himself, Lyons wanted to be sure that the Jaguars could win, so a standard 120 was taken down to Silverstone.

Lofty England takes up the story. 'We had to prove the car would go round corners, as well as go quickly in a straight line. Sir William said we would compete if an example could be driven for three hours non-stop under the lap record without breaking down. We hadn't got any racing drivers, only a couple of chaps called England and Hassan. So we did the driving. I tried to go quickly and Hassan tried to break it. Being in the experimental department, it was his job to break things. He always believes there is something wrong if he can't break it!'

Gradually they increased their speed, approaching and eventually breaking the lap record. Walter Hassan remembers the occasion well. In trying to go faster, he 'went through the bales several times and ended up with a bashed bonnet with straw all over it!' He also remembers that 'Bill Lyons said that he must try it for himself, called Rankin to join him and set off.' As they approached the first corner Lyons turned to his passenger and said 'Rankin, I've left my specs behind, indicate which way to go and thump me on the back when I must brake'. When they returned to the pits, Lyons noticed the look of fright on Rankin's face and 'laughed like hell'.

'It was decided that Hassan could prepare the cars,' recalled Lofty, 'and some old boy called England could be pit manager or something.'

For the race, Jaguar did not officially enter, but lent three cars – painted individually red, white and blue – to Leslie Johnson, Prince Bira (the blue one of course) and Peter Walker. Bira was leading when, as a result of a puncture (someone had inadvertently fitted a touring tube) he spun and could not regain the course. However, Johnson won at an average speed of 82.8 mph and Peter Walker was a good second.

Six aluminium cars were sold or lent to six selected drivers in 1950. They were works cars in all but name. Leslie Johnson had JWK 651 and it was arguably the most active of all, finishing fifth in the Mille Miglia of that year. This car along with two others was entered in the 1950 Le Mans 24-hour race merely as an experiment, because it was not expected to win, but much would be learnt about its potential, etc., that might be applied in the future. The car ran well and was in third place after 21 hours and gaining on the leaders when unfortunately the clutch succumbed. The car was retired, but valuable lessons had been learnt.

In October 1950 Johnson, with Moss sharing the driving, completed a 24-hour endurance test at the Montlhery track near Paris. The car covered the last hour at an average speed of 112.40 mph.

An important date for Jaguar and for an up-and-coming young driver called Stirling Moss, was 16th September, because it was the date of the Tourist Trophy, at Dundrod, Ulster. Tom

1

2

Wisdom, who owned JWK 688, another of the six works cars, had recognized the young driver's inherent skill and offered to lend him his XK120 for the race. Moss was quick to accept because he was keen to get some sports racing experience. Jaguar, however, were none too pleased to learn that an inexperienced youngster would be handling one of their works cars. From the drop of the flag, Johnson in JWK 651 led, but Moss quickly caught and passed him. Just before the start, the skies had opened and the track was hit by a sustained deluge. Many spun off and crashed, and it was said that the wind was blowing so strongly that it was almost a physical impossibility to steer a straight course. Rain was not unusual in this area and there was a local saying, 'If you can see the hills, it's going to rain; if you can't see the hills, it already is.'

After two hours racing, Moss was two minutes ahead of Whitehead in another XK120 who was, in turn, 27 seconds ahead of Johnson. Next came Parnell,

1 *A MK V, having started from Glasgow, descends Hucks Brow on the Shap Fell en route for Monte Carlo in 1951 (*Fox*)* **2** *The Flying Mantuan, Tazio Nuvolari, practised for the 1950 Silverstone event but sadly was too ill for the race (*National Motor Museum*)* **3** *The Moss/Johnson car high on the Monthlery banking during their 100 mph plus run for 24 hours (*National Motor Museum*)* **4** *The Whitehead/Marshall car at speed at the 1950 Le Mans finished 15th after a steady run (*J C*)* **5** *Young Stirling Moss takes the chequered flag after his quite amazing drive in the 1950 TT that earned him a works drive (*J C*)*

5

Abecassis and Macklin in relative comfort in their DB2 saloons. Shortly afterwards, the press tent blew down and the pits were constantly threatened with flooding.

After two-and-a-half hours, Moss had averaged 76.02 mph, about 98% of his target, and had been given the 'Slow' signal by his father. But, very near the end Alfred Moss, whose race records had been ruined by the weather, heard a rumour that Bob Gerard in a Frazer Nash, who had been driving through the field, had pulled up so much on handicap that he was now ahead of Stirling. So, with just one lap to go, Stirling's father hung out the 'Flat Out' sign and his son calmly responded in the increasing gloom by putting in a last lap at 77.61 mph, a Dundrod TT record.

The day after the race was Moss's 21st birthday, and as a result of this drive, Lyons invited him to drive officially for the factory the following year. When I asked him recently how he viewed the chance Wisdom gave him, Moss replied, 'Oh, that was my first big, big opportunity in motor racing because I had asked a lot of people at that time if they would let me drive their cars in the TT, which was one of the classic races of the period, but none of them would trust me. They thought I was going too fast for my experience, and if I was going to have an accident and kill myself they didn't want me to do it in their car, so none of the companies, including MG and all the much lesser companies, would let me have one, and I was very grateful to Tommy Wisdom to fix up so that I could borrow his.'

Perhaps the most famous and successful combination of car and driver ever to be seen in the world of rallying was Ian Appleyard and his XK120 Super Sports

1

2

registered and known simply as NUB 120.

Appleyard, as described earlier, had already tasted success with his SS100 – LNW 100 – and was something of an Alpine specialist. This rally in 1950 was to be his first with his new mount. He put up a fine performance and a penalty-free run, the fastest speed over the flying kilometre; and fastest time on the Col de Vars hill climb resulted in a class win and a prestigious Coupe des Alpes. This success prompted *The Motor* in its editorial to state, 'The great Jaguar Alpine performance this year raises Britain's engineering prestige in many lands.'

A first position overall followed in the Tulip Rally early in 1951, and to this were added more minor successes such as victory in the Morecambe Rally, and more importantly, another first in the RAC Rally of that year. The Alpine Rally provided a second Coupe des Alpes for another penalty-free run with his car, and entry in the London Rally ended in another best performance.

The 1951 Production Car Race at the Silverstone International Trophy Meeting showed just how popular 120s were for circuit racing. The result was a fore-

1 *Ian and Pat Appleyard and their famous XK120, NUB 120, proved virtually unbeatable in the rally world of the early 1950s (*Fox*)* **2** *Silverstone, 1951, the Production Car Race. As one might have expected, Moss walked away from the field beating the likes of Dodson, Hamilton, Wicken, Johnson and Wisdom, all in Jaguars (*J C*)* **3** *In 1951, Moss and factory man Frank Rainbow took this XK on the Mille Miglia but crashed out (*J C*)* **4** *C. Mann competing in the Eastbourne Rally in his early steel-bodied 120 Roadster (*National Motor Museum*)* **5** *Frank Grounds competing in the Morcambe Rally loses a rear spat – a not uncommon occurrence (*D. Grounds*)*

5

going conclusion. Moss finished first followed by Dodson, Hamilton, Wicken and Johnson in XKs, a DB2 finished next, followed by Holt and Wisdom in more XKs – seven of the first eight places – dominance indeed.

In 1952 Appleyard and NUB 120 completed their hat-trick in the Alpine Rally, gaining another penalty-free run. This meant another Coupe des Alpes, the third consecutive Coupe, which achievement won for the Jaguar distributor whose navigator Pat was Sir William's daughter, the first Alpine Gold Cup ever to be awarded.

Apart from NUB 120, XK120s were rallied far and wide. The French International Rallye du Soliel in 1951 was an XK benefit, with examples finishing in first, second, third, fourth and sixth places. That same year, Wood won the Scottish Rally in his 120. Later in the year, Johnny Claes and Jacques Ickx (father of the Grand Prix driver) won the notoriously tough Liège–Rome–Liège Rally in HKV 455, covering the 3,000-mile course without a single mark lost, which at the time was a unique achievement.

Domestic successes abounded, with wins in the Welsh Rally, the MCC Rally, the Rally Automobile Yorkshire and many others.

In terms of XK competition history, the big achievement of 1952 was gained by the fixed-head, LWK707. This car, the second to be produced, had been used by Bill Heynes for everyday transport. It was taken to the French circuit

1

2

3

at Montlhery in August and driven by Johnson, Fairman, Hadley and Moss for seven days and seven nights, covering 16,851 miles in 168 hours at an average speed of more than 100 mph. Such was the acclaim, that the car and drivers were given a civic reception by the Mayor of Dover on their return. The car was driven back to London by Laurence Pomeroy, Technical Editor of *The Motor*, who found everything to be in perfect order, the car betraying no signs of the marathon it has just completed.

Stirling Moss, reminded by me recently of the run recalled, 'It (the track) was very rough, but it was interesting because it was quite an achievement for that car to actually do that length of run. I must say from my point of view it was very tiring because of the monotony – no, not tiring so much as boring, because you just went round and round, and I remember one time coming round, I saw what I thought was a pretty girl. And I called in on our intercom thing, because we had a radio on it, and

1 *The third XK120 FHC to be produced had, like many XKs, a competition career (*National Motor Museum*)* **2** *The XK120 FHC crossed the line at Monthlery having covered 16,851 miles in 168 hours at more than 100 mph. Capitalizing on the achievement, the 7 days and 7 nights car is shown to the public at Earls Court in 1952 (*Fox*)* **3** *Johnson used his old faithful aluminium-bodied 120 on the 1952 RAC Rally and would have finished third but was penalized for running without the spats (*Fox*)*

I said "What's that bird like?", and they said to me, "she's terrific," so every time I went past I blew her a kiss. Unfortunately, when I stopped at the end, she turned out to be a bit of a dog, so it was all an anti-climax, but they pulled my leg on it terribly!'

The Mark VII saloons showed on innumerable occasions that, in spite of their size and weight, they came from a genuinely sporting background by winning a great many Touring Car races over a period of several years, and also by achieving good placings in various rallies.

In 1953, Ian Appleyard forsook XKs, which were no longer eligible for the Monte Carlo Rally, for a Mark VII and finished a fine second in the general classification. In his new XK, registered RUB 120, Appleyard won another Coupe des Alpes. At the International Trophy at Silverstone, the same year, Moss drove a Mark VII to victory in the Touring Car race, and at the same meeting the following year it was Appleyard's turn to lead home Rolt and Moss, all in Mark VIIs. The same year Ronnie Adams and Desmond Titterington in their Mark VII proved to be the highest placed British entry in the Monte Carlo Rally with a sixth place. Another example of the model, driven by Cecil Vard, came eighth.

Following the distinguished showing of the virtually standard XK120s in the

1

2

3

1950 Le Mans race Lyons gave the immediate go-ahead to design a serious challenger for outright honours in 1951. He was fired perhaps by the atmosphere of the event that Jaguar's had found to their surprise they had a good chance of winning or, more likely, he was confident that a specially designed Jaguar sports racing car could succeed where a production car had only just failed.

But in spite of that permission having been given while still at Le Mans in 1950, work on the new car could not begin until October of that year because the more important tasks of getting into full production the new Mark VII and the steel-bodied 120s, which had by this time replaced the earlier aluminium cars, had to take precedence.

Bearing in mind the lessons learned from 1950, and the fact that all major components had stood up well, the C-Type was to have much in common with the 120, but attention was turned to lightness, aerodynamics, braking and handling. The chassis consisted of a fully triangulated frame of round tubing of varying diameters. This frame was further stiffened by front and rear bulkheads of steel sheet welded to the frame to aid torsional rigidity. This lightweight frame was clothed with a good-looking aerodynamic aluminium shell.

The new car used the XK engine basically unchanged but with some important modifications. Larger exhaust valves ($1\frac{5}{8}$) were fitted and the exhaust porting was enlarged. Valve lift was increased by a newly designed camshaft and twin 2-in SUs were fitted to the majority of C-Types. In this form the engine gave just over 200 bhp.

The C-Type front suspension was closely related to that of the 120, with wide-based wishbones, longitudinal torsion bars and telescopic hydraulic dampers. Steering was greatly improved by the adoption of a rack-and-pinion set-up. But the rear suspension was completely new. It consisted of a Salisbury axle and a single transverse torsion bar. These were attached by a pair of radius arms further supplemented by an A bracket designed not only to provide lateral location of the axle but also to harness the torque reaction provided by fierce acceleration and use it to reduce the right hand rear wheel's habit of lifting and spinning.

Braking was improved by the adoption of a new Lockheed self-adjusting hydraulic two-leading shoe system. This was kept cool by the use of 16-inch wire wheels.

Finished just six weeks before the race, three C-Types were entered for Le Mans in 1951, driven by Moss/Fairman, Walker/Whitehead and Johnson/Biondetti. Lofty England remembers an amusing incident in practice. 'During practice Peter Walker, a very good driver indeed, came in complaining he couldn't go any faster. He had some strange tinted goggles on and I told him to put some clear ones on and he went out and broke the lap record!'

In the race Moss, who had been paired with Jack Fairman because of the latter's Le Mans experience, set a cracking pace

1 *A still hirsute Stirling Moss drove in the works' C-Type team at Le Mans in 1951* (J C) **2** *A wooden model of the C-Type used for wind-tunnel testing* (J C) **3** *A stage in the arrival at the final C-Type shape* (J C) **4** *The sorry state of PDH 33 after black ice led to an excursion and a meeting with what must have been a very solid tree* (B. Bradnack) **5** *Drivers Walker (right) and Whitehead show the relief and satisfaction of victory in the 1951 Le Mans event* (J C)

and gave the continentals a taste of things to come by repeatedly breaking the lap record with ease. But a possible 1, 2, 3 was ruined when Biondetti retired with a broken oilpipe flange. When Moss was put out with the same trouble, the Jaguar pit began to worry that the same problem might afflict the leading Whitehead and Walker car. Meanwhile, the opposition in the shape of 4½-litre Talbots, Ferraris, and 5.4-litre Cunninghams fell by the wayside, most of them defeated by Moss's early pace. So eventually Whitehead and Walker were able to cruise home victors by 67 miles. Lofty England describes the closing stages. 'Towards the end of the race the Walker/Whitehead car had built up a threequarter-of-an-hour lead which I felt was quite sufficient, and put out the SLOW sign so as to conserve the car. However, Sir William had been persuaded by somebody else that this was not sufficient and he told me to put out the FASTER sign. I did so, but in a way that Walker couldn't see it. Quite obviously he continued at the same pace and Sir William noticed this and said, "Sure he can see it, England? Better stick it out a bit further." We then had a situation whereby I was showing the FASTER signal when Sir William was looking and the SLOW one when he wasn't. Poor Walker got rather confused, so I had a word with Whitehead who was due to take over shortly. I gave him a stopwatch

1

2

3

and told him to time himself, keep to a certain time and *ignore all signals*. We hung out FASTER signs regularly, but as Sir William commented, "Not making much difference England"!'

Widespread acclaim and generous publicity were the rewards for a famous victory, and the year was completed with a second win for young Moss in the TT, driving a C-Type.

For the 1952 Mille Miglia a single C-Type was entered for Stirling Moss and Jaguar chief test driver Norman Dewis, and it was a most significant event for the two. Firstly, the car was fitted with disc brakes, which had been jointly

1 *Moss takes his second TT, driving the works C-Type that bears his favourite number (*J C*)* **2** *The development disc-braked C-Type driven in the Mille Miglia by Stirling Moss (*N. Dewis*)* **3** *The fateful long-nose C-Type that was to fail so miserably at Le Mans in 1952 (*J C*)* **4** *Duncan Hamilton, who raced Cs far and wide, negotiates the famous Goodwood chicane (*J C*)* **5** *William Heynes makes notes at Jabbeke where, in April 1953, a Mark VIII, XK120 and C-Type were run, achieving 121, 141 and 148 mph, respectively (*J C*)* **6** *The 1953 C-Types, now equipped with Webers and disc brakes, pull up outside the Restaurant des Hunundières on the Mulsanne Straight, showing that they were driven to events. The spare car is the other 17 on the right (*National Motor Museum*)*

6

developed by Jaguar and Dunlop, and secondly, stories brought home of the Mercedes' straight-line speed were to have dire effects in the near future. This event was nothing more than an experiment for Jaguar and, although lying third at one point, the car retired.

Norman, who had joined the factory as chief test driver in 1952, described to me recently the background to the disc brake development. 'When I first joined, the big push was for disc brakes, and that was my first project. I knew the people at Dunlop and we had a good tie-up. There was a good team build-up between Jaguars and Dunlop, and Mr Heynes said to me "You've got the car, get on with the development", and the first trial disc brake was put on the XK120 by Dunlop. In those days it wasn't the great big empire it is today, and there were probably only about five people involved. It was just a matter of test, test, test, until we got it right. There were problems right at the start. Nobody had any idea of what temperature the discs would run at, or the pad material, because adapting it from an aircraft brake to a car brake was an entirely different thing. You see, on the aircraft they had what we called the "wheel slide protector", where you can't lock a wheel. Well we looked at that, but it was OK on an aircraft because you'd got plenty of space. A pilot just pushed a lever on and that was it, you had maximum "decel" without locking, and of course they only do one stop and then taxi in. So really it was an entirely different requirement for the car, and temperatures were our problem for a start.

'The unit was inside the wheel and we hadn't realized what sort of airflow or air-ducting we needed to cool it, so all this was trial and error. We were seeing temperatures after a few laps of 500° and 600°C. I could finish up with the disc glowing dull red and the fluid almost bursting into flame, but it was only that way we found out and we realized we had to do something about cooling. So first we fitted wire wheels to get air from the outside, and then we had ducts in the front to bring air in on the inside of the brake, and gradually we got the temperatures down to a sensible level. We did most of our testing at Purden, a disused aerodrome just outside Wolverhampton, which Dunlop hired from the Air Ministry.

We used to go there every day, seven days a week. It was nothing to knock up five or six hundred miles a day. We used to be there at half-past-eight in the morning and stay there until it was dark at seven o'clock in the evening. We did thousands of miles that way, but it was the only way that we got it done quickly. We were ready to go to the Mille Miglia in 1952 with a brake which we knew very little about. But we thought it was the right sort of race, with the distance to be covered, so that we could see (a) just how the brake did perform and (b) what the pad life would be.

'Stirling Moss, for whom I co-drove, was briefed on the situation that we hadn't raced it before, although he knew all the work we'd done, and of course he was very interested in it, because to have something that your competitors haven't got is worth quite a lot in a race. Fortunately, we didn't strike any great problems with the brake. It was certainly a very great feature when we were racing against Caracciola in the Merc. We caught up with him and, I don't know how long we were with him, it was difficult to say, but probably about 60 or 70 miles, and we were able to outbrake him every time. He couldn't understand it at all. We could see him look every time he was "decelling" like mad, and under his braking we went by still flat out, and then put the brakes on in front of him. He said

1

2

it was incredible. Several times he didn't think the car was going to stop.'

Moss on another occasion recalling the same event, stated, 'Yes, actually the disc brakes in their early days were prone to all sorts of problems, knock-off, and vaporization, and goodness knows what else, and we learnt an awful lot with the Jaguar, with the C-Type particularly, about disc brakes. The trouble with them was that they were really powerful, but then when you used them once you couldn't use them again for quite a while until they'd cooled off, because of the vaporization of the fluid, and then you got knock-off and so on. So, because Jaguar were racing at the time, because Dunlop and Girling agreed to work together with Jaguar, that I think is the reason disc brakes came about as fast as they did.'

Le Mans 1952 is best forgotten because in Jaguar terms it was a complete disaster. The stories about the Mercedes' superiority panicked the factory into producing a more streamlined body for the C just weeks before the race. A hastily designed new cooling system to suit the lower bodywork was a complete failure and all three cars retired early on with dire-overheating.

In this period Stirling Moss attempted to persuade Lyons to build a

1 *Moss (driving) and Peter Walker took this car to second place at Le Mans in 1953 behind their team mates, after they had been delayed early on (*National Motor Museum*)* **2** *Victory, exhaustion and elation on the faces of (left to right) Angela Hamilton, Duncan Hamilton, Tony Rolt, Lois Rolt, mechancic Len Hayden and Norman Dewis (*N. Dewis*)* **3** *Le Mans victors Hamilton and Rolt do a lap of honour at the British GP at Silverstone in July 1953 (*J C*)*

Grand Prix engine. At that time the cars eligible for the World Championship were Formula 2 two-litre machines. The old XK100 four-cylinder engine was brought out of mothballs, and one example was prepared but sadly never used again.

C-Types had a fair mixture of success in 1952 and 1953 in private hands and with official entries. That this success was not more overwhelming can largely be attributed to the fact that the car was designed to win Le Mans, which was basically all that the factory seriously attempted to do.

One event at Goodwood in 1952 sticks in England's mind. 'Hamilton was leading comfortably with just a few laps to go, when suddenly he was missing. We heard he had gone off somewhere and expected to see him pottering along with a bent car. But he didn't appear. Then there was a commotion behind us and Hamilton appeared carrying a wheel, complete with drum and hub assembly. He chucked it down and announced disgustedly, "There, that's cost me 50 bottles of gin." Duncan in those days calculated everything in bottles of gin!'

For Le Mans 1953, the bodywork was back to 1951 configuration, but Weber carburettors, a lighter chassis and body, beefed-up rear suspension and, of course, disc brakes, distinguished these cars from the earlier ones mechanically. This last-named feature was by far the most important because it allowed Jaguars to set a killing pace while the Ferraris, Alfa Romeos, Cunninghams and Aston Martins had brake fade to contend with. The Cs could stop repeatedly from 140 mph plus in a shorter distance and without a trace of fade. Putting it simply, these disc brakes won the 1953 Le Mans for Jaguar and C-Types finished in first, second and fourth places.

The winning C was driven by Tony Rolt and Duncan Hamilton. Rolt, arguably one of the finest drivers Britain has ever produced, commenced his marathon experience by competing before the war in the 24-hour Grand Prix at Spa Francorchamps while he was still at Eton – indeed, he had to be driven to the event because he was too young to hold a British driving licence. Following a distinguished war career, in which he won the Military Cross and Bar for the defence of Calais and his escaping activities, including Colditz, he formed his famous partnership with Hamilton for Le Mans 1950 and 1951 in Nash Healeys, finishing fourth and sixth respectively.

After continually pestering Lofty England, Rolt was taken to Dundrod in 1951 as a reserve driver, taking over during the race the C of the unwell Johnson. He proceeded to break the lap record a number of times and worked the car up from seventh to fourth. This performance was sufficient to earn him a regular works drive for 1952. Following their earlier success together it seemed obvious that Hamilton should join Rolt in the Jaguar team for all long-distance events.

One of the last great characters of motor racing, 'Ham' drove hard and played hard. For him it was sport for sport's sake, money just did not enter into it. The Rolt/Hamilton success at Le Mans 1953 was all the more remarkable because the event featured perhaps the strongest entry ever seen, with almost every European manufacturer represented and most Grand Prix drivers competing.

Not surprisingly, this victory attracted much praise and attention. *The Daily Telegraph* described it as 'Britain's greatest motor-race triumph of all time', and Mintex headed their advertisement 'Salute to a fitting and glorious victory in Britain's Coronation Year!' *Autosport* paid tribute with a special green cover, and their editorial stated that this victory 'had gained the admiration of the whole world . . . no praise can be high enough for all concerned'.

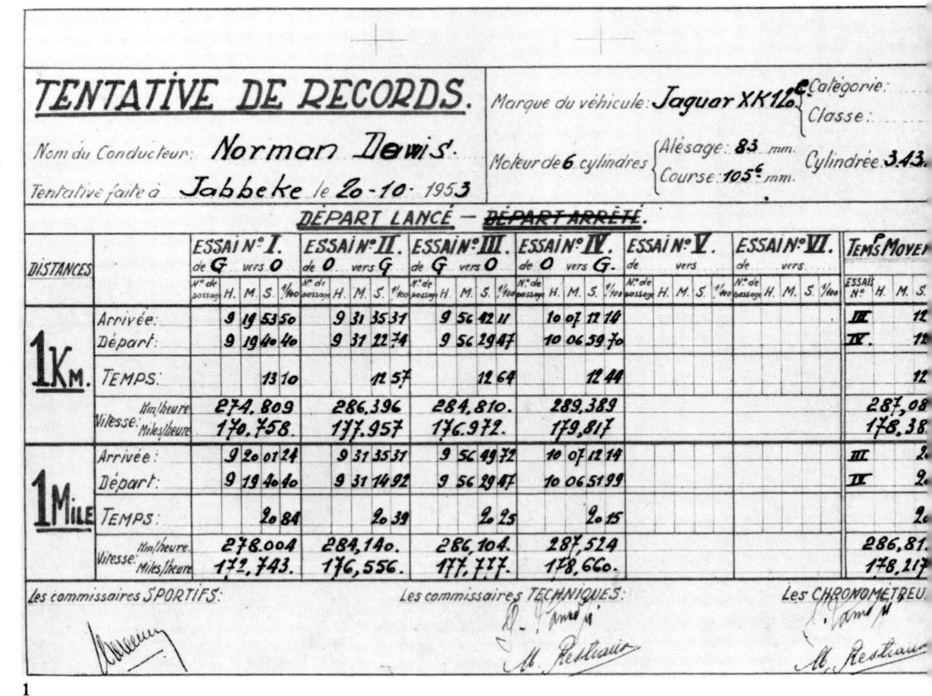

TENTATIVE DE RECORDS.

Marque du véhicule: Jaguar XK120 C — Catégorie: — Classe:

Nom du Conducteur: Norman Dewis.

Moteur de 6 cylindres — Alésage: 83 mm. — Course: 105 mm. — Cylindrée 3.43

Tentative faite à Jabbeke le 20-10-1953

DÉPART LANCÉ — ~~DÉPART ARRÊTÉ~~.

DISTANCES		ESSAI N° I. de G vers O (H. M. S. 1/100)	ESSAI N° II. de O vers G (H. M. S. 1/100)	ESSAI N° III. de G vers O (H. M. S. 1/100)	ESSAI N° IV. de O vers G. (H. M. S. 1/100)	ESSAI N° V. de vers	ESSAI N° VI. de vers	TEMPS MOYEN (ESSAI N° H. M. S.)
1 KM.	Arrivée:	9 19 53 50	9 31 35 31	9 56 42 11	10 07 12 14			III 12
	Départ:	9 19 40 40	9 31 22 74	9 56 29 47	10 06 59 70			IV. 12
	TEMPS:	13 10	12 57	12 64	12 44			12
	Vitesse: Km/heure	274.809	286.396	284.810.	289.389			287,08
	Vitesse: Miles/heure	170.758.	177.957	176.972.	179,817			178.38
1 MILE	Arrivée:	9 20 01 24	9 31 35 31	9 56 49 72	10 07 12 14			III 2
	Départ:	9 19 40 40	9 31 14 92	9 56 29 47	10 06 51 99			IV 2
	TEMPS:	20 84	20 39	20 25	20 15			2
	Vitesse: Km/heure	278.004	284,140.	286,104.	287,524			286,81.
	Vitesse: Miles/heure	172,743.	176,556.	177,777.	178,660.			178,217

Les commissaires SPORTIFS: — Les commissaires TECHNIQUES: M. Restiau — Les CHRONOMÉTREU[RS]: M. Restiau

1

2

During 1953 Jaguar produced a prototype that has come to be known in retrospect as the C/D. At the time Malcolm Sayer referred to it as the XK120C Mark II and it has also been named the Prototype D-Type, and the light alloy chassis XPII; and *Autosport* captioned a photograph of the car as 'Coventry Disco: a prototype "Disco Volante" Jaguar, one of several new types built by the Coventry concern for sports car races.' Because the last C-Type was numbered XKC 053, this car was designated XKC 054. It was quite literally a stepping stone from the C-Type to the D-Type that was to be. It was of tubular frame construction but did not have such good aerodynamic properties as the D. It suffered from having a greater frontal area, presumably because it embodied the wet-sump C-Type engine.

The C/D made only two semi-public appearances. One was at Jabbeke, where Norman Dewis took an XK120, fitted with undershield and bubble top, and proceeded to achieve the quite remarkable speed of 172.412 mph.

The C/D that achieved 178.383 mph was therefore not so impressive, although it may have had a problem with the fuel system because the car not pulling well at the top end. Norman Dewis told me recently that 'in comparison with the XK it was struggling at the beginning of the measured mile.'

The second and last appearance of the car was at the 1954 Le Mans Test Weekend, where the factory also had the

1 *The results of the C/D's Jabbeke runs, officially tabulated* (J C) **2** *Hamilton tries the C/D at Silverstone* (N. Dewis) **3** *Dewis at speed with the many-named prototype, which was so reminiscent of the later E-Type and a certain mid-engined car* (N. Dewis) **4** *The entire Jabbeke entourage surrounds the specially prepared XK120. Malcolm Sayer, wearing the jacket, leaning on the XK's bonnet* (N. Dewis)

prototype D. The C/D was immediately overshadowed by the D, with its better aerodynamic qualities and, as a result, development was concentrated on the latter.

Although the C/D was forgotten and eventually broken up, it was not without significance in the evolution of the Jaguar sports racing car. What is particularly interesting is that from most angles it looked rather like the E-Type of eight years later.

As previously mentioned, Norman Dewis took an XK120 'to get some publicity before the Motor Show', according to England, to the familiar ground of the Jabbeke motorway. There were aerodynamic improvements but relatively little mechanical modification. 'Don't go too fast, Norman,' instructed Lofty. 'Take it a bit steady first time'. Dewis takes up the story. 'We thought we would probably see 161/162, which would have been ample, and Malcolm Sayer said he wouldn't like to visualize it going any quicker, because it would just go airborne, and I still don't know how it stayed on the road now! I think fortunately for me, there was not a breath of wind. I am afraid if there had been possibly a gust blowing on the car at threequarters angle, I think it might have started pushing it off. The rev limit I was told from the work they had done on the test bed was five-five/five-six – that should be alright. Well, I had already got up to six before I got to the measured mile, you see, so I thought, well, I can't back off now, I must keep this going. It was beautiful, there was no problem, so I just held it there and got six-two all the way through.

'It did wander a bit. It was a dual carriageway but although you've got two lanes, it doesn't look very wide at that speed. It did weave a bit but I think that was because it was going a bit light on the front end and the rear was probably steering it a bit, that was all. We only had one problem. It was a good job I did a trial run on the day before. We had a cockpit bubble which was screwed down on top of me, and I went straight down Jabbeke just trying it. I'd got around 80/90, then 100, 120, 130, 140, and there was traffic on the road, and I was so engrossed looking at the instruments, the rev counter, the oil temperature, water and all that, that I'd gone a bit farther than I had anticipated, at least 12 or 13 miles. I then cut across the grass centre and went down the other side, and it was halfway down the other side, I started to feel a bit hazy, and I suddenly realized I'd got no air coming in at all, just heat and fumes. I thought, God, I can't open this canopy, I felt I'd got to open it, but I thought, if I push my fist through it that's going to ruin it, so I just gulped as much air as I could and held my breath and got back. I was virtually collapsing and when they opened it they said I looked like a red tomato! We remedied it by cutting some small louvres in the canopy, but it was just as well we had the trial run and found out, otherwise it might have spoilt the proper runs.'

At the completion of the runs England phoned Lyons who immediately asked, 'How did you get on?' 'Not very well sir,' replied England, '172'. 'Yes, yes, but what's that in mph?'

To keep the competition chaps on their toes, Lyons would occasionally produce a mock-up of his own in conjunction with Fred Gardner who was superintendent of the sawmill. One such animal was the Brontosaurus, as it was nicknamed, a veritable monster.

Dewis recalls, 'Gardner would never let anyone in his shop no matter who you were, and we knew something was going on because Sir William was popping

1

2

down there in the evenings about five o'clock and staying till around nine. Then we heard they were building a car and all sorts of rumours were going round that it was going to be a new sports car with a small engine'. The experimental shop then made it into a runner and Dewis was asked to test it. On asking Sir William what it was intended for, he received the reply, 'It's not for racing but it might offer something for record-breaking'. However, it was only ever run two or three times at Gaydon on their new two-mile runway'.

Around this time Jaguar thoughts again turned to Grand Prix racing. For 1954 the engine limit was to be 2½ litres, and a 2.4 version of the XK engine was developed with this in mind. Furthermore, two or more C-types were cannibalized to build a single-seater. Basically, two C frames were cut in half and the front halves welded back to back. Various all-enveloping single-seat bodies were drawn and one mock-up built, but the project proceeded no further because the sports cars were gaining all the publicity required.

The XK140 did not enjoy the same dramatic competition success as the XK120 for several reasons. Competitors had by now had some time to catch up and both racing and rallying were begin-

1 *Norman Dewis, looking a little apprehensive, prepares for the run that was to amaze all concerned* **2** *Revs clocking 6200 and a mean speed of 172 mph would be impressive today, let alone in 1953.* (N. Dewis) **3, 4** *The Brontosaurus showed that William Lyons was human if nothing else. Streamlined it was, but the front wheels could not steer!* (J C)

ning to develop into the more professional science that they have become today. The factory, though never officially, prepared XK120s for selected entrants; but to the best of my knowledge, they never prepared any 140s to any great extent. They had intended to have a trio of rally-prepared 140 Roadsters but, sadly, the uncertainty surrounding the future of a number of international events following the Le Mans disaster caused the abandonment of these plans.

But a number of private enthusiasts entered XK140s in diverse events around the world with a fair share of success. One such competitor was none other than Ian Appleyard who had replaced his second 120 Roadster with VUB 140, a special equipment 140 FHC. Although officially retired, he entered a few domestic events with the car, and in 1956 came second overall in the RAC Rally, which was no mean feat. In the same year, Guyot won his class in a 140 Roadster in the Mille Miglia.

In fact, 120s continued to clock up successes in a variety of events. In May 1954, Gillie Tyrer presumably in a Fixed Head, finished second in a Closed Car Race at Ibsley. In July, Haddon and Vivian continued the model's success in the International Alpine Rally with a win. The same month Lord Louth took his class at the Bouley Bay hillclimb in Jersey, and a month earlier Mike Salmon had similar success in the Bo'ness Speed Trials. In August Bob Berry finished third behind Michael Head (father of present Williams Grand Prix designer Patrick Head) in a C-Type at Snetterton, and came second in the O'Boyle Trophy Handicap Race at The Curragh. And still in August, a trio of 120s occupied the first three places in a production car race in Ohio.

The legendary D-Type, successor to the C-Type and evolved from the C/D prototype, was first seen at the private Le Mans test session early in 1954. In just three laps (two more than the officials had given permission for) Rolt proceeded to break the previous year's lap record, set up by Ascari's Ferrari, by five seconds. Such was a foretaste of things to come, of a racing career that was to last more than six years and would include innumerable wins all over the world including three at Le Mans – the race the car was designed to win.

Jaguar and Sir William, as has already been stated, knew the immense publicity value to be gained from success at Les Vingt-Quatre Heures du Mans, which considerably outweighed that gained from any other event. Le Mans put a manufacturer's name on the map worldwide.

The Sarthe circuit of more than five miles per lap and an average lap speed of more than 100 mph demanded a high top speed, good acceleration and consistently powerful braking rather than particularly outstanding roadholding. The D-Type was designed with this in mind, and explains why the car was not always so competitive, particularly in later years, over shorter, twistier circuits.

Central to the D-Type's design was its new monocoque centre section. This 'tub', constructed of 18-gauge magnesium alloy, gave strength tempered with lightness. To the front bulkhead was argon-arc-welded a sub-frame of round and square aluminium tubing which carried the front suspension, steering, engine and ancillaries. To the rear, a double-skinned bulkhead was attached to the live Salisbury rear axle by four trailing arms, two above and two below the axle. Springing was provided by a transverse torsion bar and location by an A bracket. Front suspension was basically similar to the XK's, with top and bottom wishbones and longitudinal

1

2

orsion bars. Also bolted to the rear ulkhead was the tail section, which was nstressed and carried the two flexible uel tanks and spare wheel. Dunlop disc rakes were fitted all round, and newly leveloped Dunlop light alloy wheels of ither 16- or 17-inch diameter and $5\frac{1}{2}$-nch width.

The engine, basically the same XK nit with three Weber DCO3 45 mm arbs, was mounted at 8° from the verti-al and produced 250 bhp at 6000 rpm. The C-Type head was further developed vith the enlarging of the inlet valves to $\frac{7}{8}$-inch diameter, and slightly different camshafts were used. The gearbox was entirely new, being Jaguar designed and all synchromesh. Particularly significant was the fact that the engine now had dry sump lubrication, which reduced the sump depth by half and allowed a lower bonnet line and a better centre of gravity. Axle ratios ranged from 4.09:1 to 2.53:1, the latter making possible a top speed in the region of 200 mph.

The superb body styling was the work of Jaguar's aerodynamicist Malcolm Sayer, and was the result of extensive wind-tunnel testing with one-tenth scale models. Surely a perfect example of the

1 *A typically active, privately campaigned XK 120 belonging to Midlands businessman Bertie Bradnack, seen here with a young Bob Berry at the wheel. Berry later took over from Rankin as head of publicity (*B. Bradnack*)* **2** *Bertie Bradnack rounding the bottom 'S' at Shelsley (*B. Bradnack*)* **3** *Practice for the Rheims 12-hour event in which Hamilton (seen here) and Rolt finished second behind Whitehead and Wharton (*National Motor Museum*)* **4** *OKV 2 was the third D-Type built and had been driven by Moss and Walker at Le Mans (*N. Dewis*)* **5** *Duncan Hamilton pitting for new goggles in the closing stages of his and Rolt's titanic struggle to catch the leading Ferrari during the very wet 1954 Le Mans race (*J C*)*

5

'if it looks right, it is right' school of thought, the D-Type had a one-piece, forward tilting, easily removable alloy bonnet, a streamlined head fairing that concealed the fuel filler, and a distinctive fin fitted on the competition cars for straight-line stability.

Curiously, Jaguar's new sports racing car did not gain its name immediately, with the first four cars having XKC chassis numbers (XKC 401–404) and subsequently XKD numbers (XKD 405 and 406 completed the Ds that were built in 1954).

The D-Type's debut at Le Mans in 1954 should have been a glorious one but instead it was a mysterious one. Sabotage has been suggested, but whether this is true or not the fact is that the three D-Types, and only the D-Types, had misfiring problems which were eventually traced to the presence of a fine grey sand in the fuel. Rolt and Hamilton, delayed early on, battled valiantly only to be delayed again when Rolt was forced into a sandbank by another competitor. In spite of this and torrential rain (Hamilton was getting wheelspin at 170 mph), they provided one of the most thrilling finishes ever seen at the French circuit ending up just 105 seconds behind the winning Ferrari of Gonzales and Trintignant.

A little later the tables were turned at the Rheims 12-Hours when Moss destroyed the Ferrari opposition before retiring. Whitehead and Wharton went on to win and Rolt and Hamilton finished second to give Jaguar a fine one-two.

Referring to 1955, Lofty England recalls, 'We managed to acquire the services of one, Hawthorn. A lot more testing was done, more serious preparations carried out and we even flew the cars out to Le Mans.' For 1955, several revised D-Types were built, four of which were retained by the factory (XKD 504/5/6/8) and one was prepared for Briggs Cunningham (XKD 507). Subtle but important changes distinguished these cars from the 1954 cars. For example, XKD 501/2 (both supplied to Ecurie Ecosse at the beginning of 1955) and XKD 503 (similarly supplied to the Belgian national team) and the other 1955 cars had a simplified front sub-frame which was now bolted to the front bulkhead, instead of being welded, to enable easier and speedier repairs. The radiator and oil cooler were mounted on a small subsidiary sub-frame for similar reasons and, most noticeable, the bonnet was lengthened by $7\frac{1}{2}$ inches to improve penetration. The 'long-nose' cars, as they came to be known, had their wrap-around screens extended back as far as the head fairing to combat buffeting from side winds. More importantly, there were improvements to the cylinder heads in the constant search for more power. The inlet valve diameter was increased to 2 in and the exhaust valve from $1\frac{5}{8}$ in to $1\frac{11}{16}$ in. To avoid valve overlap, the inlet valves were retained at their previous 35° inclination, but that of the exhaust valves was increased to 40°, hence this is known as the 35/40 head or perhaps more generally as the 'wide-angle' head. Power output as a result of these modifications was raised to 275 bhp at 5750 rpm.

Le Mans 1955, saw the D-Type win, but it was not a victory in which to rejoice because it was the year of the awful crash. It is sad quite apart from the obvious reasons, because the race had the makings of a titanic struggle. Until the accident, Fangio in the Mercedes 300SLR and Hawthorn in the long-nose D had been having a tremendous duel during which Hawthorn set the lap record at 122.39 mph. Neck-and-neck they raced round the long fast circuit of

1

2

the Sarthe, swopping places, bringing the crowd to their feet and sending the reporters into ecstasies. Some have likened it to their famous duel in the French Grand Prix. The Mercedes were withdrawn early on Sunday morning and Hawthorn and Bueb went on to a rather hollow victory.

At the TT at Dundrod the Jaguar/Mercedes battle was renewed but it ended for Hawthorn conclusively when his crankshaft broke. 'We didn't think the car suitable for the TT at Dunrod in Ireland,' recalled England, 'so we just entered one car. We had Hawthorn and Titterington, who lived in Belfast, which kept the expenses down! We were worried about tyre wear for, at Dundrod in 1953, tyres lasted only 35 miles. However, Dunlop had developed a new Stabilio tyre which was steel-braced, and this was much more successful. Anyway, with Moss in the Merc and Hawthorn in the D we were having a good dice when we called Mike in for fuel and a change of tyres. However, to our surprise the tyres were hardly worn, and we sent him out again without

1 *Hectic activity during a D-Type night refuelling stop* (N. Dewis) **2** *Friendly rivals, Moss drove in the team from 1951 to 1954 and Hawthorn in 1955 and 1956* (J C) **3** *From very amateur beginnings the famous Scottish team, Ecurie Ecosse, became an increasing force to be reckoned with* (N. Baldwin) **4** *Grounds hustles his car along in superb scenery during the 1956 Monte, in which another Mark VII, that of Adams, finished first* (D. Grounds) **5** *The 1955 works cars lined up at Le Mans with Lofty England masterminding affairs. The new longer nose is clearly visible* (J C)

5

changing them. A couple of laps later a Mercedes mechanic was holding a sign calling Moss in, when Neubauer rushed forward and bodily picked him up and removed him. Two laps later Moss crashed when a tyre blew. We were very surprised that the normally so ultra-efficient Germans should make a mistake and I mentioned this to chief engineer Uhlenhaut when I met him some months later. He said they hadn't made a mistake, but when we didn't change our tyres they thought their calculations must be wrong!'

The year 1955 saw a class win and fourth overall in the Tulip Rally for the Mark VII, and at the Daily Express meeting Mike Hawthorn led the field to notch up another win. The Monte of that year included a works team of cars driven by Adams, Vard and Appleyard and, although success largely eluded them individually, they collectively won the Charles Farroux Team Trophy. Later in the year a Mark VII won its class in the RAC Rally.

A works team again contested the Monte early in 1956 and this time their reward was outright victory by Adams in his faithful Mark VII. This was perhaps the highspot of the model's illustrious competition career. The year 1955 saw notable private entries such as those of Ecurie Ecosse and Duncan Hamilton score wins in D-Types in more minor events around the globe.

Six new long-nose cars were built for 1956. These differed in various small details – the most noticeable being the provision of a full-width screen to meet a change in the regulations for Le Mans. Early events provided a series of retirements and failures rather than successes, but morale was boosted with factory D-Types finishing in first, second, and

1

third places at Rheims.

'At Rheims,' recollects Lofty England, 'we had the famous incident when Hamilton ignored my pit signals. He came up to me afterwards and said, "S'pose I'm in trouble?" "No trouble at all," I replied, "You just don't drive for us any more!" Some months later at Christmas he sent me a card and parcel. Opening it I found it contained a mortar board and cane. Anyway, some time later he was due at the factory to collect a D he had bought and I told the gate police to tell me immediately he arrived. I knew he would come straight up to my office and I would have about 30 seconds warning. I got a phone call to say he had arrived. I put the mortar board on, took the cane in my hand and sat down menacingly at my desk. At that moment there was a knock on my door. At my summons it opened and a commissionaire walked in. He seemed to find the situation amusing and started to buckle at the knees. "What's so funny, man. Go and fetch Mr. Hamilton." When he returned with Duncan, there were a number of heads craning to try to see in through the door!'

Le Mans 1956 was a disaster for the Jaguar factory but not for the Jaguar marque. Of the three factory cars, two crashed on the second lap and the third was badly delayed with a split fuel line. But Ecurie Ecosse, those famous privateers from Scotland, came to the rescue and Flockhart and Sanderson drove their D-Type to a fine victory.

The XK140 had even less success in racing than in rallying for here a trend had been started some years previously towards specially designed sports racing cars rather than specially prepared production sports cars. One of the main propagators of this had been Jaguar themselves with the C-Type, and even more so with the D-Type.

However, a lone XK140 FHC entered at Le Mans in 1956 far from disgraced itself. The car, a virtually standard Special Equipment model with some 25,000 miles recorded, was driven by a pair of comparatively inexperienced drivers, Peter Bolton and Bob Walshaw. The car was prepared by the factory at a cost of just £1,200 and set off from home with all tools, spares and supplies on the roof. At the halfway stage the XK was lying a most creditable 14th, and as the hours passed, moved as high as 12th, beating through reliability and performance, cars with far greater racing pretensions. But with several hours to go the car was called in by the officials and disqualified for refuelling a lap earlier than was allowed some hours previously. This was bitter disappointment but at least it is on record that the XK140 Fixed Head covered 1,749 miles at an average speed of 83 mph. However, I recently discovered in conversation with a later owner of the car, that the French had made a mistake. Bolton and Walshaw should never have been disqualified and the French apologized profusely. The car they probably should have disqualified? None other than the winning D-Type.

Mike Hawthorn, works driver in 1955 and 1956, seems to have a special place in Jaguar history because everybody at

1 *The lone 140 FHC of Bolton and Walshaw that did so well until cruelly disqualified at Le Mans in 1956* (National Motor Museum) **2** *Mike Hawthorn in familiar surroundings with the famous visor which he preferred to goggles* **3** *Frank Grounds exercising the suspension of his Mark 1 saloon during the 1956 RAC Rally* (D. Grounds) **4** *In 1957 Ecurie Ecosse acquired 1956 works cars. Jack Fairman takes one of the cars through Becketts at Silverstone* (J. Fairman)

the factory who knew him remembers him with special affection.

Mike gave the new 3.4 saloon its first victory at the Daily Express meeting at Silverstone in September 1957, being chased by his great friend, and frequent partner-in-crime behind the scenes, Duncan Hamilton. He had several more wins in these cars, racing his own modified road 3.4. Ironically and very tragically he was killed in this car on the Guildford by-pass at the beginning of 1959, having been world champion the previous year and having just announced his retirement from motor racing.

Meanwhile in October 1956, the factory issued the following press release: 'The information gained as a result of the highly successful racing programme which the company has undertaken in the past five years has been of the utmost value, and much of the knowledge derived from racing experience has been applied to the development of the company's products.

'Nevertheless, an annual racing programme imposes a very heavy burden on the technical and research branch of the engineering division, which is already fully extended in implementing plans for the further development of Jaguar cars.

'Although withdrawal from direct participation in racing in the immediate future will afford much needed relief to the technical and research branch, development work on competition cars will not be entirely discontinued, but whether the company will resume its racing activities in 1958 or whether such resumption will be further deferred must depend on circumstances.'

Ironically, 1957 was Jaguar's best year at Le Mans, with the five D-Types entered finishing in first, second, fourth and sixth places. Ecurie Ecosse again provided the winning car, driven by Flockhart and Bueb, and also the second car piloted by Sanderson and Lawrence. The French pair Lucas and Mary were third, the Belgians Frere and Rousselle fourth, and Hamilton and Gregory, although in the fastest car of all, were delayed by the exhaust burning a hole in the floor and had to be content with sixth.

Later in 1957 Ecurie Ecosse were involved in a curious event when the Scottish team were invited to take part in a contest organized at Monza and entitled the Race of Two Worlds. The theory was that it should be a race between a number of American Indianapolis cars and sports cars from Europe over a distance of 500 miles. In practice the event was boycotted by the newly formed International Professional Driver's Union which excluded all the Europeans bar the Scots, who would have nothing of any boycott.

The D-Types stood no chance of success on the banked bowl against eight Indianapolis cars designed precisely for this form of track. The Ds were restricted to a maximum of 160 mph by problems with tyre temperature after Jack Fairman in practice recalled, 'Just as I was approaching the banking at about 165 mph, the entire right hand back tread came off with a noise like a six-inch shell'.

Because of the bumpy nature of the track, it was decided to run the event in three heats of 166 miles with an hour's break between each. The American cars gradually fell apart until only three were left. The Ds ran like clockwork and finished a very creditable 4th, 5th and 6th, driven by Fairman (who acquired his nickname of Fearless Jack from the event), John Lawrence and Ninian Sanderson, respectively.

This was the last D-Type success of real importance and, not surprisingly after four years of racing, they became less competitive with Lister-Jaguars taking over and a three-litre engine

1

2

being mandatory for Le Mans. Such an engine was produced but it never proved reliable.

The private owners and teams continued as in the previous years to clock up innumerable victories and included among them a young farmer from Scotland named Jim Clark, who drove the Border Reivers D-Type to a number of successes. Clark's career is inseparable from motor racing history and his later achievements are well known, but what is perhaps not so familiar is his early involvement with Jaguars.

Clark's first proper season was with that other Scottish team, the Border Reivers, a group of wealthy young farmers. He drove the team's by then ageing D-Type, TFK 9, and it was during that season that he achieved the first 100 mph lap by a sports car on an unbanked track in this country, and that his team manager Ian Scott-Watson was

1 *The year 1957 brought a second victory for Ecurie Ecosse at Le Mans with this D-Type heading the Jaguar domination. Flockhart is in the car with Wilkie Wilkinson on the door and co-driver Bueb on the rear wing (*D. Grounds*)* **2** *Wilkie Wilkinson, the Ecurie Ecosse chief mechanic, stands by Jack Fairman awaiting the start of the 1957 Monzanapolis event* **3** *Fairman completed the 500 miles to finish fourth at an average of 150 mph – probably the highest average speed in any European event up to that time (*J. Fairman*)* **4** *The most successful of all Jaguar-engined sports racing cars was the Lister, seen here in what has come to be known as the 'knobbly' body guise* **5** *D-Types were still notching up successes in the late 1950s and Jack Fairman corners OKV2, the third D to be built, in a full four-wheel drift at an event in 1958 (*J. Fairman*)*

5

moved to mention in correspondence with Lofty England that they had a young driver in their team of 'extreme promise'.

He later drove the 'flat-iron' Lister Jaguar for the same team, gaining useful international experience.

Having built sports cars with MG, Bristol and Maserati engines, with mixed fortunes, Brian Lister was persuaded to copy one of his customers and fit an XK unit. This was the beginning of a successful partnership that was to dominate sports car racing for several years to come in one form or another.

The basis of Lister's design was a chassis formed by two large 3-inch tube side-members and these were cross-braced. Front suspension was by unequal-length wishbones and coil springs, and the rear by a de Dion axle located by four trailing arms and again coil springs. One of the features of a Lister was that the entire body or section thereof could be removed easily in minutes, which must have been appreciated by mechanics.

The new car first appeared in March 1957 piloted by that marvellous driver Archie Scott-Brown, whose name was inseparably linked to that of Lister until his untimely death. Unfortunately, the car had clutch trouble, which ruined any chance of winning, but clocking the fastest lap was some reward. The next meeting produced a win in the British Empire Trophy race and the combination went on to 11 more wins and a second out of the 14 races entered that season.

Replicas of this ultra-successful car began to be built in small numbers in 1958 at a cost of £2,750 plus purchase tax, and customers included Ecurie Ecosse and Briggs Cunningham, who was to have a tremendous amount of success over the next couple of seasons in the US with his Listers, driven mainly

1

2

3

4

by Walt Hansgen.

A single-seater Lister was built to special order for Ecurie Ecosse to campaign in a re-run of the previous year's Monzanapolis event. Jack Fairman drove the car but it broke a valve in the second heat.

In 1958 the wins continued to accumulate, with two Lister works cars, MVE 303 and VPP 9, being driven by Scott-Brown (until his tragic death mid-season), Moss, Hansgen and Bueb. This body style, which has come to be known as the 'knobbly' Lister, was superseded by a new body designed by aerodynamicist Frank Costin, and these rather more bulbous cars have come to be known as the Costin bodies.

A few successes came in 1959 but nothing like so many as in the previous two seasons. So, with the gradual emergence of the Cooper and the Lotus rear-engined cars and the lack of competitiveness of the 3-litre version of the XK engine, Brian Lister decided to call it a day at the end of the year.

It would take a complete book to describe all the competition successes of the Mark 1 and 2 models, the racing successes in the UK, in Europe, in other continents such as Australia, the rallying successes everywhere and the records set up. Perhaps the first of the very long line of triumphs was the Spa Production Touring Car Race in May 1956. Paul

1 *Silverstone 1958, and Mike Hawthorn (VDU 881) and Tommy Sopwith (EN 400) entertain the crowds in their Mark 1 saloons* (J C) **2** *Hawthorn and Sopwith cornering hard during their splendid duel* (J C) **3** *The XK150 was never a serious competition contender but allowed keen club drivers such as Don Smith to have plenty of fun* (D. Smith) **4** *Bernard Consten climbs Mount Ventoux on his way to victory in the 1961 Tour de France, a success he had first achieved in 1960 and one he was to repeat in Mark 2s in 1962 and 1963* (J C)

Frere, with a modified 2.4, proceeded to win his class and put up best performance irrespective of class, having lapped at 97 mph. The 3.4s took over the mantle of Touring Car supremacy from the Mark VIII, being lighter, and when the Mark 2 3.8s became available, that supremacy was further confirmed.

A long list of successes alone does not make for interesting reading, but a list of some of the more distinguished drivers who have competed in Mark 1s and 2s might. In roughly chronological order they were: Duncan Hamilton, Mike Hawthorn, TEM 'Tommy' Sopwith (who formed a team of Jaguars under the title of Equipe Endeavour), Sir Gawaine Baillie, Ron Flockhart, Walt Hansgen, Ivor Bueb, Roy Salvadori, Jack Sears, Stirling Moss, the late Graham Hill, the late Colin Chapman, Peter Sargent, the late Bruce McLaren, the late Mike Parkes, the late Peter Lindner, Mike Salmon, Mike MacDowel and Denny Hulme. Some of the best motor racing seen on television was provided in the early 1960s by the dicing between Graham Hill, Roy Salvadori and Mike Parkes.

In rallying many successes were gained, notably the Morley brothers' class win in the Tulip Rally of 1958 in their 2.4; the winning of the Charles Farroux Team Trophy in the Monte of 1959 by a team of 3.4s; a win overall for the Morleys in the Tulip Rally of the same year, this time in a 3.4; class wins in the 1960 Alpine and RAC rallies; and successive class wins in the Tour de France, including first, second and third places in 1962.

In May 1961, the company was able to advertise: 'For the tenth year, Jaguar wins Silverstone Production Touring Car Race, taking first, second, third, fourth, fifth (all 3.8s) and seventh (2.4) places'. In smaller print it continued, 'since the inauguration in 1949 by the BRDC and the *Daily Express* of the series of International Trophy Meetings at Silverstone, Jaguar cars have achieved

1

2

3

1, 2 *E2A at the factory, just before being taken to Le Mans where it raced in Cunningham's colours. Driven by Walt Hangen and Dan Gurney, it was the fastest in practice but suffered a series of mechanical troubles in the race and finally gave up the ghost after nearly 10 hours (*J C*)* **3** *The E2A was fitted with a 3.8 engine (note the power bulge) and shipped to the USA, where Cunningham raced it with mixed fortunes (*J C*)* **4, 5** *The E-Type's first race: Roy Salvadori in the Coombs car, BUY 1, and Graham Hill in the Equipe Endeavour car, ECD 400, leave the pack behind (*J C*)*

SUPER-PLUS
SUPER
SHELL
visco-static
SUPER-PLUS
DAILY EXPRESS
visco-static
SUNDAY EXPRESS
MINTEX
MINTEX
DUNLOP
ECD 400
17 TVX
89

36
ECD 400

13 consecutive victories in the 13 annual meetings held there including all 10 Production Touring Car Races held annually since 1952'.

In January 1964, a list of the previous year's successes, mostly gained with Mark 2s, necessitated a two-page spread and included Endurance Events (wins in the *Motor* International Six-Hour Touring Car Races, etc.), Long-Distance Records, International Rallies and Touring Car Races (in Australia, New Zealand, Tasmania, the UK and Germany).

The long-distance records were set up by a team of five drivers led by Geoff Duke at Monza in a 3.8 Mark 2 and International Class C (3000–5000 cc): 10,000 miles at 106.58 mph; 15,000 km at 106.61 mph; 3 days at 107.02 mph; and 4 days at 106.62 mph.

As the company stated in their press release, retirement from racing was only to be temporary. In fact, the factory fire, an over full order book, and the need to devote all available time to developing new replacement production models meant that Jaguar did not return to racing. Nevertheless, they kept an eye on developments at Le Mans, just in case. Sir William was quoted as stating, 'The company has certainly not lost interest in racing. We have not neglected the necessary development work for returning to the sport. We have made certain plans but just when we will put them

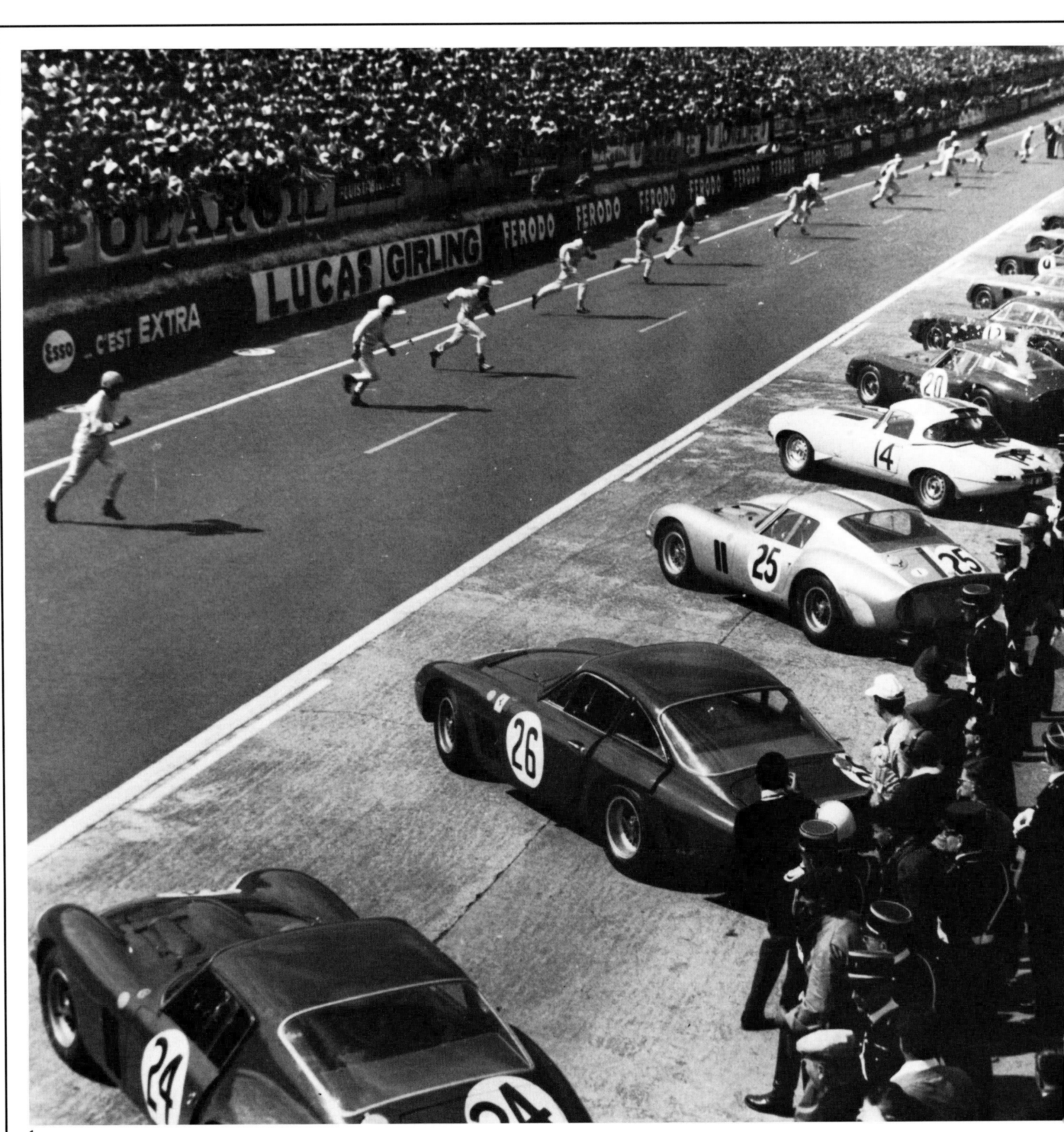

1

nto operation must depend on circumstances'.

The second development E-Type, E2A (A for aluminium) was the logical development of the D-Type, and it had been intended to run it at Le Mans in 1958. Its major difference from the Ds was its independent rear suspension. Later, the car was prised out of Jaguar by Briggs Cunningham and raced by him at Le Mans in 1960. Although it set the fastest time in practice, it lasted only a few hours in the race, its 3-litre engine letting it down.

The factory subsequently swopped the 3-litre unit for a 3.8, and the bonnet acquired a power bulge. In this form the Cunningham team campaigned the car briefly with mixed fortunes. Walt Hansgen achieved one victory with the car, but Jack Brabham could not repeat the result when he drove it.

The E-Type's racing debut was a fairy-tale one, because a few weeks after the car had been announced, two completely standard cars were taken to Oulton Park for a 25-lap GT Trophy race, where Graham Hill and Roy Salvadori proceeded to finish first and third respectively against opposition that included Aston Martin DB4 GTs and 250 GT Ferraris.

Three E-Types were privately entered for Le Mans in 1962 and while they were not serious contenders for outright honours, by dint of reliable running they finished fourth and fifth, driven respectively by Salvadori/Cunningham and the Peters, Lumsden and Sargent.

In spite of this and other early successes, the E-Type could not really compete with the race-bred Ferraris, so some gradual development of the cars took place. Notably, the John Coombs semi-works car, originally registered BUY 1 when raced by Salvadori, but later changed to 4WPD, came in for constant attention and modification. A new monocoque of thinner gauge steel was substituted and aluminium bonnet was fitted. The engine was modified and the suspension gradually developed. The car raced in this guise in 1962 and, in spite of a considerable amount of testing by Hill, although competitive, it was never outstandingly successful.

It was clear that the Ferrari 250 GTOs had a considerable weight advantage, and that if Jaguar wanted the E to be competitive, that was one of the main areas for attention. As a result, in late 1962 the factory started to build a few competition E-Types (they have come to be known more widely as lightweight Es) for the coming season. The main difference was the use of an all-aluminium centre monocoque and aluminium block for the engine, which also had fuel injection and produced more power than any other XK unit up to that time. The engine also featured dry sump lubrication and the D-Type wide-angle head. The lightening exercise resulted in a saving of some 500 lbs over the standard car

2

3

4

1 *The Competition (lightweight) E-Type pitted against the might of Ferrari at Le Mans in 1963 (*J C*)* **2** *Robin Sturgess, one of the first regularly to race an E-Type, at Silverstone. This car, the 12th to be built, later appeared in the film* The Italian Job, *registered 848 CRY (*National Motor Museum*)* **3** *This car, one of the trio entered at Le Mans in 1963, was the most successful, finishing in ninth place after a variety of adventures (*J C*)* **4** *In an effort to improve the drag factor of the E-Type, Sayer produced this coupé, which Dick Protheroe acquired and used to good effect (*J C*)*

and approximately 100 lbs over the Ferraris.

These Lightweights, of which about 12 were produced, proceeded to be raced far and wide, including three Briggs Cunningham cars entered at Le Mans. Two of the cars retired but the one registered 5115 WK finished in 9th position. However, apart from a few domestic triumphs, real success again eluded the E-Type drivers, as they came into contention more and more with thinly disguised sports racing cars designed purely with racing in mind.

It is a little sad that these competition modifications were not made earlier in the car's life when it might have covered itself in rather more glory, but the factory at that time had neither the time nor the interest to devote itself wholeheartedly to racing.

Three-times world champion Jackie Stewart had strong Jaguar connections in the early part of his career. Brother Jimmy raced C-Types for Ecurie Ecosse and the two brothers were partners in the garage business started by their father. Dumbuck Motors in Dumbartonshire were Jaguar area dealers and someone once remarked that the arrangement was ideal – ie, Jackie bends them and Jimmy mends them!

In 1962–63, Jackie drove Jimmy's E-Type and said later that it was 'the most forgiving motorcar I have driven . . . you could make it do anything you liked'. As a result of his driving of the E, David Murray picked him to drive the Ecurie Ecosse Tojeiro and Cooper Monaco. This was the beginning of his climb to the heights of Formula One. On the way, he drove Eric Brown's famous XK120 registered 1ALL, and tested and raced the Coombs lightweight E.

Not surprisingly, the trusty XK engine has appealed to a number of specialist builders both for road cars and for racing machines, but mainly for the latter. Sprint specialist Gordon Parker had a series of specials that were Jaguar-powered, notably the Jaguette, fitted with a 1939 SS Jaguar $2\frac{1}{2}$-litre ohv engine; the Jaguara with XK unit; and finally the HK Jaguar with twin superchargers feeding its XK engine. Frank Le Gallais, from Jersey, produced a rear-engined device that had a succession of power units including a $3\frac{1}{2}$-litre Mark 5 unit and ultimately a tuned XK engine.

Oscar Moore fitted an XK unit with great success in his HWM, and the factory followed suit with their sports cars. Another famous name who produced a handful of similarly powered sports racing cars was Cooper. These Cooper

1

2

3

Jaguars, although never outstandingly successful, gained a number of places. Apart from the Listers, dealt with already, the other XK-engined sports car worth noting was the Tojeiro Jaguar, produced in 1956.

In addition to those mentioned, there have been many others, such as Rivers Fletcher's single-seater, Mike Barker's Alton Jaguar, Col Rixon Bucknell's special, the Hansgen Jaguar Special, the Galaxie-engined E-Type called the Egal, various production Panthers in the 1970s, the RGS Atalanta, the RRA Jaguar of Geoff Richardson, the Emeryson-Jaguar and others.

XK engines have enjoyed great success not only on land but also on water. Best known of the record-breaking protagonists was Norman Buckley, who started with an early unit in 1949 adapted for marine use and fitted in his Ventnor Hull. He took the one-hour, three-hour and twenty-hour nautical records. In 1951 with a modified engine, *Miss Windermere II* achieved a mean of 79 mph, and in 1953 and 1954 Buckley, with a different hull, graduated to a C-Type and then to a full three-Webered Le Mans spec. unit.

In 1954, the German von Mayenburg captured the world's eight-hour record with his XK-engined craft. And in 1958 Buckley with his engine now up to D-Type spec, wrested the one-hour record from the German at a speed of 89.08 mph, and with further development was achieving more than 120 mph by 1959. The boat's hull was damaged when dropped from a hoist, so *Miss Windermere IV* was built, this one being a prop-riding hydroplane. Powered by a virtually standard engine from Buckley's 1962 E-Type, a speed of 118.75 mph was achieved in 1970.

Apart from record-breaking boats,

1 *The ultimate sports racing car built by Jaguar, the XJ13, at high speed on the MIRA banking* (N. Dewis) **2** *The heart of the XJ13, its magnificent four overhead cam V12 engine set amidships* (N. Dewis) **3** *There's no mistaking the parentage – the XJ13 was the culmination of the C/D, D-Type, E2A, E-Type theme* (N. Dewis) **4** *Norman Dewis takes the XJ13 out for another run on the MIRA track, the scene of much Jaguar testing over the years* (N. Dewis) **5, 6** *Dewis, who started with XK120s developing disc brakes, nearly finished with the XJ13. Miraculously, the incredibly experienced Dewis escaped relatively unscathed from this monumental shunt, which was probably caused by tyre deflation* (J C)

there have been many well known off-shore racing power boats. John Coombs's boat, named *Cheetah*, was fitted with two 3.8-litre engines. *Tramontana II* and *Jackie S* each boasted a total capacity of 15.2 litres and an output of more than 1000 bhp, with four 3.8 units apiece. The former was owned by Richard Wilkins and driven by Jeffrey Quill, the latter owned and driven by financier Dr Emil Savundra.

The next logical step in Jaguar sports racing car design after E2A was the mid-engined XJ13, the initials standing for 'Experimental Jaguar'. Aware that more power than the six-cylinder engine could yield would be needed for racing in the 1960s, Jaguar's chose the V12 configuration. The image boost that such an engine would engender was a desirable asset for its eventual use in the production range.

The engine was a four overhead cam unit with a capacity of a fraction under 5 litres. Inevitably in this form it looked rather like two XK heads with a common crankcase. Power output was 502 bhp at 7600 rpm, and it had dry sump lubrication.

Unlike the earlier D and E types, the XJ13 was fully monocoque in construction. It consisted of two main sill boxes joined by the floor, and a double bulkhead at the front with a single one at the rear. To this rear bulkhead was bolted the engine, which was stress-bearing and carried the transaxle and rear suspension. At the front, the suspension and water and oil radiators were carried by another aluminium box section. The suspension was basically E-Type and the gearbox from ZF.

The bodywork was the work of aerodynamicist Malcolm Sayer and a continuation of the beautiful D-Type theme. Originally, it was planned as a closed GT car and we can but conjecture that as such it would have been stagger-

1

2

3

ingly good looking. Although the XJ13 was designed during 1965, it was not completed until 1966 and was not really tested until 1967.

For a variety of reasons Jaguar did not want the press to know that they had a possible Le Mans challenger, and Sir William gave orders that the car was not to be circuit-tested. This eventually became too much for Norman Dewis, who discussed it with Lofty England. Lofty in turn said that if Dewis wanted to take it quietly to MIRA one Sunday he knew nothing about it! This Dewis did, and several days later was summoned to Sir William's office. The conversation, according to Norman, went something like this:

Sir William: 'I thought I gave orders that the XJ13 was not to be tested.'
ND: 'Yes you did, sir.'
Sir William: 'And yet you took it to MIRA.'
ND: 'Yes, I did sir.'
Sir William: 'When I give an order I expect it to be obeyed. Don't ever disobey me again.'
ND: 'Yessir.'
Sir William: 'Well man, how did it go?'

During 1967, David Hobbs tested the XJ13 round MIRA and set the unofficial record for the fastest lap at any circuit in Britain, namely, 161.6 mph.

Some time later, a tyre blew out while Dewis was going round MIRA and the car somersaulted, luckily without serious injury to the driver, but with considerable injury to the car. Fortunately, Abbey Panels still had the

1 Bob Tullius in the Group 44 prepared V12 E-Type leading one of its main rivals, which he continually defeated in spite of being outnumbered (J C) 2 Schenken and Rouse lead from the start of the 1977 TT at Silverstone, but as usual it was not to last (J C) 3 The Group 44 V12 XJS in Trans-Am action with 1979 champion Bob Tullius driving (J C)

body formers, so the car could be rebuilt.

On reflection, it is very sad that the XJ13 was not developed more rapidly because many feel that it could have beaten the GT40s and regained for Jaguar the supremacy they had enjoyed in the 1950s.

With regard to body styling, the late Malcolm Sayer's influence on the Jaguar line was second only to that of 'the old man'. He joined the company as a trained aerodynamicist and was responsible for the C-Type, the development C/D prototype and his most famous design, the D-Type. From the original car he evolved the famous long nose and fins of various shapes and sizes. He must have been one of the first to apply aircraft principles to sports racing car design based on his close connections with the Royal Aircraft Establishment at Farnborough, and their wind tunnel.

Sir William liked to work with a full-scale mock-up, but Sayer preferred to draw his cars full-scale on the wall, working out every detail in theory first. Although the influence of his designs for a road-going mid-engined car can be seen in the current XJS, his *pièce de résistance* must surely be the XJ13, the climax of the Jaguar racing line and whose styling he was responsible for.

In mid-1974 Bob Tullius of Group 44 persuaded BL that the ageing V12E could be raced successfully in Category B sports car events in the US, and that this would give a fillip to dropping sales. BL agreed and commissioned Group 44 of Virginia and Huffaker Engineering of California to cover their respective coasts. The results were impressive. Huffaker won first time and Group 44 finished the season, which for them only started in August, as Northeast Division Champions with five victories and seven track records from seven starts.

The following year was similarly successful, with ten starts netting seven victories and eight track records, enabling Tullius to become SCCS National Champion.

The racing XJ Coupés were built and raced in 1976 and 1977, just two seasons in which they at times showed their fantastic potential, but sadly never realized it by winning any event. Some say that they were not given sufficient time to prove themselves. We shall never know.

The cars, prepared to FIA Group 2 regulations applicable to the European Touring Car Championship, were fitted with 5416 cc fuel-injected V12 engines developing more than 530 bhp. Initially

1

they used a wet sump lubrication but later changed to a dry sump system. Rather controversially, the cars were built and prepared not by the factory but by Broadspeed, independent of the factory.

The first year, 1976, was an embarrassing one for all concerned, because British Leyland, who had announced the car with a fanfare of publicity long before it was properly developed, had very red faces when the cars repeatedly failed to start.

The following year was more successful, with a second place at Nürburgring and pole position at every event bar one. Time and again the Leylands (not Jaguars) proved to be considerably quicker than their main rivals the BMWs, but they just could not last the distance at the pace.

For a while the Coupés provided a great spectacle, making what would otherwise have been a dull championship more exciting. It is sad that they could not have earned just a little glory. Derek Bell, one of the regular team drivers, told me recently, 'I enjoyed the cars, and they could have been very good if they had been persevered with a little longer'.

In the States, luckily things went very much better. Towards the end of 1976 Group 44 switched to an XJS and won a Sports Car Club of America race at Lime Rock. The engine in this car produced 475 bhp aided by no less than six Webers and dry sump lubrication.

In 1977 Group 44 entered the Trans Am Championship and won five out of ten Category One races, beating Porsches and Chevrolet Corvettes in the process, with the V12 now developing 536 bhp. This gave the car a 0–60 acceleration time of 5 seconds, with 100 mph coming up in a further 5.3 seconds, and a top speed of around 190 mph. Weight was reduced by 240 lbs by fitting a fibreglass bonnet and plexi-glass rear window. These wins plus a second and third gave Tullius and Group 44 the Category One championship. The following year was similarly successful, with Tullius and Group 44 taking the driver's and manufacturer's titles.

1 *'Jaguar back in racing' ran the headlines, and what better sequel than 'Jaguar wins at Donington'? (*J C*)* **2** *After a fine drive by Martin Brundle in awful conditions, this was the view seen by the opposition at the end of the Donington race (*J C*)*

In 1982 the highly respected race car preparation company, Tom Walkinshow Racing (TWR), prepared an XJS for racing in Australia and in doing so had the idea that it would make an ideal basis for tackling the BMWs in the European Touring Car Championship, Group A sector. With backing from Motul, a French oil firm, and the approval of Jaguar, TWR set out to do just that. Entering the fray mid-season they bagged four wins, including a first and second at the TT at Silverstone. This was the first Jaguar victory in this most historic event since Moss's win in a C-Type in 1951.

Impressed by the professionalism of the team and the unquestionable success gained, Jaguar felt able to come out in the open and officially back the team as official entrants. Thus Jaguar were back in racing (XJCs had been entered as Leylands or whatever, not as Jaguars) after 26 years.

The philosophy of Group A is that a car's major areas must in principle be unaltered from their production counterparts. Bodywork must remain unaltered and suspension principles must be adhered to, with the aim of proving the inherent ability, or lack of same, of a car's design, rather than to accept thinly disguised racing cars that bear no relation to the standard product.

The TWR XJS was given an engine developing 400 bhp as opposed to the standard 299. Walkinshaw commented that they could get 500 bhp but the consumption would be a problem, because the extra fuel stops would cancel out any speed advantage. With a top speed in the region of 170 mph and a weight of almost $1\frac{1}{2}$ tons, braking was also a potential problem. To overcome this, 13-inch ventilated discs were fitted all round, those at the rear being mounted outbound.

In 1983 the results came gradually. In the first race at Monza, one of the two team cars finished second by $3\frac{1}{2}$ seconds after the bonnet had come loose and a stop had to be made to tape it down. A third followed at Vallelunga. At Donington, young Martin Brundle put in a stirring drive in heavy rain to snatch a superb victory. Further victories were gained and at the year end the score stood at BMW 6, Jaguar 5, a worthy feat when one considers that the two XJs were taking on more than a dozen BMWs.

Meanwhile in the USA, Group 44 had not been resting on their laurels. The

1

XJs had been further developed, particularly in the engine department, and had gained three Trans Am wins in 1981. But these engine developments were intended for something even more exciting.

Jaguar, now with its new enlightened leadership, had expressed a desire to compete in a world series rather than in a purely domestic one. Hence the idea of building a purpose-designed prototype with the Jaguar V12 engine powering it. Nobody would admit it although it was hinted at, but Jaguar's old stomping ground – where a certain 24-hour event took place in France – was at the back of their minds.

As a result, a striking GT prototype was designed by American Lee Dykestra for Tullius. It was an aluminium Honeycomb monocoque design with steel bulkheads and the V12 mounted amidships. The fibreglass bodywork enclosed a design along ground-effect principles then current in Formula One. The car was initially intended for the American IMSA series, but was built with long-distance endurance events in mind. The XJR-5, as it was known, gained several wins in the USA in 1983 and in 1984 it was decided to enter Le Mans, more as an experiment than a serious attempt to win. With Brian Redman now a team member and John Watson joining for Le Mans, the cars far from disgraced themselves against the sheer weight of Porsches.

For 1984 TWR added a third car plus former ETC champion Hans Heyer and proceeded to dominate the Series. By round 6, the halfway mark, they had amassed 5 wins and on two occasions a 1st, 2nd and 3rd. Such was the new enthusiasm for competition that Sales and Marketing Director, Neil Johnson, stated that sales had risen as a result and that racing had 'greatly contributed to our reputation for quality. Jaguar's name,' he concluded, 'can only be enhanced by a professionally run racing programme'.

1 *The XJR-5 underwent considerable development in 1982 and 1983 – the few races entered were regarded as tests only (*J C*)* **2** *The stylish V12-engined XJR-5. Driving in 1982 and 1983 was shared by Tullius and Canadian Bill Adam (*J C*)* **3** *The XJR-5 certainly looks very different from a D-Type but will it emulate its famous ancestor? (*J C*)*

2

3

10 STYLING THEMES

The 'Lyons line' is a famous phrase in motoring circles and is confirmation that this first-rate businessman also influenced the design of most of his products very heavily. Also, we should not forget the part that the highly respected aerodynamicist Malcolm Sayer played in literally helping to shape the range by applying principles learnt in competition design and development.

1

Before World War II, William Lyons tended to follow contemporary fashion and produced cars with long low bonnets and Bentley-type styling. But after the war it was another matter. Lyons led fashion, 'daring', as the adverts put it, 'to be boldly individual'. But where did his influences come from? It is interesting to note the similarity between certain models of Bugatti and later Jaguars. This should not be taken as criticism of Lyons's work, but rather as a hint as to where part of his inspiration may have been derived.

The Type 57 Bugatti in two of its many forms provides a couple of interesting comparisons. The T57S in Atalante guise has many similarities to the XK120 Fixed-head of approximately 20 years later. The relative positioning of headlamps, sidelights, grille and bonnet profile are very similar, and in side view silhouette the 120 is a development of Jean Bugatti's theme – first seen as a Lyons theme in the one-off SS100 fixed head.

For competition purposes the T57S chassis was fitted with a streamlined, all-

1 *The T57 Bugatti was produced in a number of forms but this body, called the Atalante, was particularly striking and must have seemed very modern in the mid-1930s (*National Motor Museum*)* **2** *The one-off SS100 Coupé, shown at the 1938 Earls Court Show, was Lyons' version of a developing theme among performance car designers (*J C*)* **3** *Jumping 15 years ahead, one can surely detect the same theme in the XK120 Fixed-Head Coupé, further refined (*J C*)* **4** *A one-off built in 1946 or 1947 shows the XK120 Roadster taking shape in Lyons' thinking*

4

enveloping body termed the 'tank'. Notable amongst the car's successes was its triumph in the 1937 Le Mans race, when the car established a new record distance. It does not take very much imagination to notice the similarity of the 'tank' and the C-Type designed for Le Mans, particularly the ill-fated long-nose, long-tail version.

Again, and perhaps most interestingly, the Bugatti influence (or was it coincidence?) can be seen when you compare the type 101 with several Jaguar models. This T101 was to all intents and purposes a re-bodied, postwar version of the T57 and was announced in 1951, some four years after Ettore Bugatti's death. The obvious Jaguar comparison is with the 2.4 saloon, later to be augmented by the 3.4, and later still to be replaced by the MK2 range. The similarities are too obvious to need comment.

What is less well known is that during 1954, Lyons had designed a new body for the replacement of the XK120, and it was only lack of time and staff that resulted in the 120's survival in revised form as the XK140. The prototype model was quite obviously another step in the evolution of the 'Lyons line' and can also be compared with interest.

The Bugatti/Jaguar relationship is further compounded by the fact that the original Bugatti-designed Bebe Peugeot is said to have been the inspiration for

1

2

3

the Austin Seven, which was Lyons's original basis.

Whatever the real sources of Lyons's inspiration, it all makes for fascinating conjecture. It may also stimulate discussion on the subject among enthusiasts, because there are plenty of them all over the world treasuring examples of the unforgettable 'Lyons line'.

1 *This T57 45 Bugatti 'tank' car appeared for practice at Le Mans in 1937 but was not raced (*Geoffrey Goddard Collection) **2** *The better known and successful 'tank' T57 Bugattis showed an early exercise in streamlining (*National Motor Museum*)* **3** *That the later C-Type Jaguar should be developed with the aid of wind-tunnel testing says a lot for the earlier Bugatti design (*N. Dewis*)* **4** *The post-war T57, re-bodied and shown as the T101, compares interestingly with a number of Jaguar models (*National Motor Museum*)* **5** *A very odd photographic mock-up, which was an attempt to create a four-seater sports car by stretching a 120 FHC – something done in less drastic fashion later for the 140 (*J C*)* **6** *A one-off pure mock-up that was considered as a replacement for the XK120, but practical considerations at the factory, which was battling to meet demand, precluded the production of a completely new model. The 2.4 shape is visible at the front, to be announced some time later (*J C*)*

6

1, 2, 3 *Three fascinating views of a styling exercise in open and closed form. Produced probably in the mid-1950s, it may have been intended as an XK140 replacement, instead of the XK150. In it can be seen the small saloon, the XKs and even a hint of the XJC. It was a very stylish car, aided in appearance by the lack of sidelights, number plates, etc. Note the XK120 front bumpers*

1

2

3

AFTERWORD

by

John Egan

Chief Executive of Jaguar Cars Ltd

Whenever I read an account of Jaguar's illustrious history – such as the one produced by Philip Porter – I am reminded of just how much all of us at Jaguar Cars today owe to Sir William Lyons and his team for the legacy of excellence they established over the years.

Sir William not only founded the company but bequeathed to its products all the qualities for which they are now famous – beautiful styling, fine performance, excellent handling, value for money and a competition pedigree that rivals that of any car manufacturer in the world. I am convinced that all these factors helped the company survive when Jaguar's fortunes reached their nadir in the late 1970s.

The recovery that has taken place at Jaguar over the past four years since the reconstitution of the company in 1980 has been based on the hard work and dedication of all employees from the shopfloor to the boardroom. By working together as a team we have proved that the British Motor Industry can design, manufacture and sell luxury cars to the highest international standards. More than anything else, though, we have recognised the sovereignty of the most important person in our business – the customer.

Jaguar Cars Ltd will face many challenges in the future but provided we strive continually to satisfy customers I am confident we can succeed.

John Egan
January 1984

INDEX

Abecassis, George, 118
Adams, Ronnie, 122, 136
Alpine Trial, 112, 115, 119, 132, 142
Alvis, 14, 80
Appleyard, Ian, 63, 81, 115, 116, 118, 119, 120, 122, 132
Ascari, 132
Aston Martin, 76, 145
Attwood, Richard, 81
Austin Seven, 9, 10, 157
Austin, Sir Herbert, 9, 14
Autosport, 63, 70, 128, 129

Bailey, Claude, 47, 48
Baillie, Sir Gawaine, 142
Barker, Mike, 100, 147
Bell, Derek, 151
Benbow, Gordon, 100
Bentley, 30, 53, 70, 99, 155
Berry, Bob, 132
Biondetti, C., 123, 126
Bira, Prince, 116
Birmingham Evening Mail, 107
Black, Captain, 42
Bloomfield Street, 7, 8
Blue Star Garages, 81
BMC, 94
BMW, 48, 76, 105, 151, 152
Border Reivers, 139
Bourgeois, Madame, 44
Brabham, Jack, 145
Bristol, 140
British Motor Holdings, 94
Brooklands, 35, 112, 114
Brough Superior, 9, 19, 80
Brown and Mallalieu, 12, 81
Browns Lane, 52, 102
Bueb, Ivor, 138, 141, 142
Bugatti, 47, 53, 155, 156

Campbell, Sir Malcolm, 31
Carraciola, Rudolf, 126
Chapman, Colin, 142
Cheltenham Motor Club, 60
Chrysler Plymouth, 22
Citröen, 45
Claes, Johnny, 120
Clark, Jim, 139
Clyno, 14
Cocker Street, 9
Coombes, John, 81, 145, 146, 148
Cooper, 141
Coventry, 13, 39
Coventry Climax, 8, 108
Coventry Eagle, 9
Crisp, Trevor, 109
Cunningham, 124
Cunningham, Briggs, 84, 134, 140, 145, 146

Daily Mail, 90
Daily Telegraph, 128
Daimler, 8, 104
Dawes, Nigel, 62
Dennis, 47, 80
Dewis, Norman, 60, 94, 125, 129, 130, 131, 149
Dodson, 120
Donington, 114
Dot, 9
Dundrod, 68, 116, 118, 135
Dunlop, 44, 50, 71, 74, 100, 108, 126, 127, 133

Ecurie Ecosse, 136, 137, 138, 140, 141, 146
Edinburgh, Duke of, 67
Edwardes, Sir Michael, 107
Egan, John, 107
Elizabeth II, Queen, 67
England, F. R. W. "Lofty", 50, 104, 105, 111, 116, 123, 124, 125, 128, 130, 134, 135, 137, 140, 149
Equipe Nationale Belge, 62
ERA, 35
Essex, 22
Evans, P. J., 81

Fairman, Jack, 121, 123, 138, 141
Fangio, 134
Ferrari, 54, 74, 124, 132, 134, 145, 146
Fiat, 14
Flockhart, Ron, 137, 138, 142
Follett, Charles, 81
Frazer Nash, 118
Frere, Paul, 138, 142
Frey, Emil, 11

Gardner, Fred, 130
Gardner, Major Goldie, 48
Geneva Motor Show, 53, 83
Gerard, Bob, 118
Goddard, "Jumbo", 70
Gonzales, 134
Goodwood, 128
Gordon Crosby, F., 30
Gregory, Masten, 138
Group 44, 150, 151
Guy, 8

Hadley, Bert, 121
Hamilton, Duncan, 70, 120, 128, 134, 136, 137, 138, 142
Hansgen, Walt, 141, 142, 145
Harriman, Sir George, 94
Harrop, Jack, 112, 114, 115
Hassan, Wally, 35, 47, 48, 108, 109, 114, 116
Hawthorn, Mike, 134, 135, 136, 137, 138
Helliwell Group, 41
Henlys, 12, 17, 24, 31, 81
Hetherington, Mrs. V. E. M., 114
Heynes, William, 25, 30, 32, 34, 45, 47, 48, 100, 114, 120, 126
Hill, Graham, 142, 145
Hillman Aero Minx, 22
Hillman Wizard, 22
Hindes, B. W. J., 63
Hobbs, David, 149
Holbrook Lane, 13, 22, 52
Holmes, Denis, 90
Holt, 120
Huffaker Engineering, 150
Hulme, Denny, 142
Humber, 25
H. W. Motors, 81

Ickx, Jacques, 120

Jabbeke, 8, 50, 60, 116, 129, 130
Jacob, E. H., 112
Johnson, Leslie, 48, 116, 117, 120, 121, 123
Johnson, Neil, 153

Kennings, 81
Knight, Bob, 107

Lanchester, 22
Lawrence, J., 138
Le Mans, 47, 60, 61, 62, 67, 74, 84, 111, 116, 123, 127, 128, 129, 132, 134, 136, 137, 138, 144, 145, 149, 153, 156
Lewis, Hon Brian, 112
Lex Garages, 81
Liège–Rome–Liège Rally, 120
Lindner, Peter, 142
Lister, Brian, 140
London Rally, 119
Lotus, 141
Lumsden, Peter, 145

McLaren, B., 142
MacDowel, M., 142
Macklin, Lance, 118
Mann Egerton, 81
Mansbridge, Reg, 81
Marcus, Sidney, 81
Maserati, 140
May, Michael, 108, 109
MCC Rally, 120
Meadows, 8
Mercedes Benz, 76, 105, 126, 134, 135, 136
Midland Motor Museum, 100
Mille Miglia, 116, 125, 126, 132
MIRA, 62, 149
MG, 10, 118, 140
Monte Carlo Rally, 67, 115, 122, 136
Montlhery, 116, 121
Morecombe Rally, 119
Morley Brothers, 142
Morris Cowley, 10
Morris Isis, 22
Morris Minor, 17
Moss, Alfred, 118
Moss, Stirling, 54, 116, 118, 120, 121, 122, 123, 124, 125, 126, 127, 134, 135, 136, 141, 142, 152
Motor, The, 8, 119, 121, 144
Motor Panels, 40
Mundy, Harry, 108, 109

Naish House, 115
Neubauer, A., 136
Newsome, Sammy, 81, 112, 114, 115
Norton, W. C. N., 115
Norwegian Viking Rally, 64

Owen, H. R., 81

Paris–Nice Trial, 114
Parkers, 12
Parkes, Mike, 142
Parnell, Reg, 117
Perak, H.H. The Sultan of, 14
Pomeroy, Laurence, 121
Prescott, 115
Protheroe, Dick, 81

RAC Rally, 112, 114, 115, 119, 132, 136, 142
Radford, Harold, 81
Radford Works, 102
Randle, Jim, 109
Rankin, Bill, 30, 112, 116
Redman, Brian, 153
Renault, 22
Rheims, 67, 134, 137
Rhode, 22
Ritchie, James, 14, 81
Road and Track, 112
Robinson, Geoffrey, 105, 107
Rolls Royce, 53, 70, 99, 107
Rolt, Tony, 122, 128, 132, 134
Rossleigh, 81
Rothwell and Milbourne, 81
Rover Speed Pilot, 22
Rover Ten, 22
Rubery Owen, 30, 40

Salmon, Mike, 142
Salvadori, Roy, 142, 145
Sanderson, N., 137, 138
Sargent, Peter, 142, 145
Sayer, Malcolm, 129, 130, 133, 148, 150
Scott-Brown, Archie, 81, 140, 141
Sears, Jack, 142
Shelsley Walsh, 112, 114, 115
Silverstone, 8, 116, 119, 122, 142, 152
Smith, Rowland, 81
Society of Motor Manufacturers, 51
Sopwith, Tommy, 142
SS Car Club, 27, 114
Standard Big Nine, 14
Standard Enfield, 15
Standard Little Nine, 22
Standard Motor Co., 14, 19, 30, 42, 44
Star, 22
Stewart, Jackie, 81, 146
Stewart, Jimmy, 81, 146
Sturgess, Robin, 81
Sunday Times, The, 108
Sutton, R. M. V., (Soapy), 50, 51
Swallow Doretti, 41
Swallow Gadabout, 41
Swift Ten, 14, 15

Talbot, 124
Taylor, Bob, 112, 114, 115
Thomson and Taylor, 35
Titterington, Desmond, 122, 135
Tom Walkinshaw Racing, 152, 153
Tour de France, 142
Tourist Trophy (T.T.), 68, 116, 118, 135, 152
Trintignant, M., 134
Triumph TR2, 41
Trojan, 22
Tube Investments, 41
Tulip Rally, 116, 119, 136, 142
Tullius, Bob, 150, 151, 153

Uhlenhaut, R., 136

Vanden Plas, 44, 104

Walker, Peter, 116, 123, 124
Walmsley, William, 7, 8, 9, 14, 16,
Walter, Martin, 81
Watson, John, 153
Watson, Miss V., 115
Watsonian, 41
Welsh Rally, 112, 114, 115, 120
Weslake, Harry, 25, 30, 32, 48, 78
Wharton, Ken, 134
Whitehead, Peter, 117, 123, 124, 134
Wicken, 120
Wisdom, Tom, 34, 35, 112, 114, 117, 118, 120
Wolsley Hornet, 16, 17

FREDERICK WARNE
Penguin Books Ltd, Harmondsworth, Middlesex, England
Viking Penguin Inc., 40 West 23 Street, New York, New York 10010, U.S.A.
Penguin Books Australia Ltd, Ringwood, Victoria, Australia
Penguin Books Canada Ltd, 2801 John Street, Markham, Ontario, Canada L3R 1B4
Penguin Books (N.Z.) Ltd, 182–190 Wairau Road, Auckland 10, New Zealand

First published in 1984

ISBN 0-7232-3248-2

Printed in Great Britain by BAS Printers Limited, Over Wallop, Hampshire